Local Government Innovativeness in China

Local government innovation has become one of the most important topics on China's policy agenda in recent decades. This book explains why some local governments are more innovative than others.

This book uses a novel theoretical framework and points out that in China's multi-level government structure, the administrative hierarchy and the span of control could shape local governments' innovation motivation, innovation capability, and innovation opportunity, thus influencing local government innovativeness. The author systematically analysed the 177 winners and finalists of the biennial Innovations and Excellence in Chinese Local Governance (IECLG) Awards Programme from 2001 to 2015 to provide convincing empirical evidence to support this theory.

This book adopts an institutional approach to explaining local government innovativeness in China and may be a useful reference to help us learn more about local government decisions and behaviours.

Youlang Zhang, PhD, is Assistant Professor in the School of Public Administration and Policy and Research Scientist in the Beijing Academy of Development and Strategy at Renmin University of China. His research interests include policy process, citizen participation, and bureaucratic politics. His recent publications have appeared in *Public Administration Review, Journal of Public Administration Research and Theory, Public Administration, Policy Studies Journal, American Review of Public Administration, Public Management Review, International Public Management Journal, China Quarterly*, and other journals.

Routledge Focus on Public Governance in Asia
Series Editors:
Hong Liu, *Nanyang Technological University, Singapore*
Wenxuan Yu, *Xiamen University, China*

Focusing on new governance challenges, practices, and experiences in and about a globalising Asia, particularly East Asia and Southeast Asia, this focus series invites upcoming and established researchers all over the world to succinctly and comprehensively discuss important public administration and policy themes such as government administrative reform, public budgeting reform, government crisis management, public-private partnership, science and technology policy, technology-enabled public service delivery, public health and aging, talent management, and anticorruption across Asian countries. The book series presents compact and concise content under 50,000 words long which has significant theoretical contributions to the governance theory with an Asian perspective and practical implications for administration and policy reform and innovation.

Political Economic Perspectives of China's Belt and Road Initiative
Reshaping Regional Integration
Christian Ploberger

Exploring Public-Private Partnerships in Singapore
The Success-Failure Continuum
Soojin Kim and Kai Xiang Kwa

Collaborative Governance of Local Governments in China
Jing Cui

Local Government Innovativeness in China
Youlang Zhang

For more information about this series, please visit www.routledge.com/Routledge-Focus-on-Public-Governance-in-Asia/book-series/RFPGA

Local Government Innovativeness in China

Youlang Zhang

LONDON AND NEW YORK

First published 2021
by Routledge
2 Park Square, Milton Park, Abingdon, Oxon OX14 4RN

and by Routledge
52 Vanderbilt Avenue, New York, NY 10017

Routledge is an imprint of the Taylor & Francis Group, an informa business

© 2021 Youlang Zhang

The right of Youlang Zhang to be identified as the author of this work has been asserted in accordance with sections 77 and 78 of the Copyright, Designs and Patents Act 1988.

All rights reserved. No part of this book may be reprinted or reproduced or utilised in any form or by any electronic, mechanical, or other means, now known or hereafter invented, including photocopying and recording, or in any information storage or retrieval system, without permission in writing from the publishers.

Trademark notice: Product or corporate names may be trademarks or registered trademarks, and are used only for identification and explanation without intent to infringe.

British Library Cataloguing-in-Publication Data
A catalogue record for this book is available from the British Library

Library of Congress Cataloging-in-Publication Data
A catalog record has been requested for this book

ISBN: 9780367460839 (hbk)
ISBN: 9780367636487 (pbk)
ISBN: 9781003026808 (ebk)

Typeset in Times New Roman
by Deanta Global Publishing Services, Chennai, India

Contents

List of figures	vii
List of tables	viii
Preface	ix
Abbreviations	xiii

1 Introduction   1

2 Explanations for local government innovation in China   7

3 An institutional theory of local government innovativeness   22

The structural characteristics of the multi-level bureaucratic organisations in China   22
Local government innovativeness along the administrative hierarchy   24
Local government innovativeness along the span of control   27

4 The Innovations and Excellence in Chinese Local Governance Awards Programme   35

5 An empirical investigation based on the IECLG Awards Programme   45

The measurement of local government innovativeness   45
The administrative hierarchy and local government innovativeness   46

The span of control and local government innovativeness 50
Provincial-level statistical analysis 53
City-level statistical analysis 58

6 Conclusion   65

Appendix   73
Index   93

Figures

2.1	The frequency of "government innovation" found in sources printed between 1960 and 2019 in Google's text corpora (Google Books Ngram Viewer)	8
2.2	Existing explanations for local government innovation adoption in China	9
3.1	The hierarchy of the Chinese government system	23
3.2	The relationship between administrative hierarchy and innovation resources	25
3.3	The relationship between administrative hierarchy and innovation motivation	26
3.4	The relationship between administrative hierarchy and government innovativeness	27
3.5	The number of prefecture-level regions governed by each province in China	28
3.6	The number of county-level regions governed by each prefecture-level city in China	29
3.7	The theoretical framework	31
4.1	The total number of IECLG winners and finalists (2001–2016) in each province in China	38
5.1	The correlational relationship between the number of city-level IECLG winners and finalists and the number of city-level regions in each province from 2001 to 2015	51
5.2	The correlational relationship between the number of county-level IECLG winners and finalists and the number of county-level regions in each prefecture-level city from 2001 to 2015	52

Tables

2.1	An overview of some explanatory empirical studies published in the mainstream disciplinary journals of PA and PP (listed by publication year)	11
4.1	Description of the applicants, finalists, and winners of the IECLG Awards Programme	37
4.2	A list of the number of IECLG winners and finalists by province and year (2001–2015)	39
4.3	The specific initiators of IECLG winners and finalists (2001–2015)	41
4.4	The typology and distribution of the IECLG winners and finalists (2001–2015)	43
5.1	The number of IECLG winners and finalists across government tiers and years	48
5.2	The typology and distribution of the IECLG winners and finalists across government tiers	49
5.3	Variables, measures, and data sources for provincial-level analysis (2001–2015)	54
5.4	Summary statistics of the variables for provincial-level analysis (2001–2015)	56
5.5	Provincial-level analysis results (2001–2015)	58
5.6	Variables, measures, and data sources for city-level analysis (2005–2015)	59
5.7	Summary statistics for city-level analysis (2005-2015)	61
5.8	City-level analysis results (2005–2015)	62
A.1	A list of all winners and finalists in the IECLG Awards Programme (2001–2015)	74

Preface

This book adopts an institutional approach to explaining local government innovativeness, the extent to which a government is capable of generating innovations, in China. Since the 1960s, extensive studies in the disciplines of economics, management science, political science, sociology, public administration, and public policy have examined government innovations. Nevertheless, it was not until the 2010s that scholars attempted to systematically use empirical data to analyse why and how local government innovations occur in China. Local government innovation has become one of the most important topics on China's policy agenda in recent decades. Since the reform and opening up in 1978, China's central, provincial, and city governments have continuously expressed support for local government innovations, thereby providing scholars and practitioners with an excellent opportunity to examine local government innovation dynamics in China.

This book first provides a systematic review of previous studies on local government innovation in China. Existing explanations for local government innovation in China mainly focus on internal determinants (e.g., macro-level jurisdictional and governmental characteristics or micro-level policymakers or policy entrepreneurs) and external pressures (e.g., vertical top-down or bottom-up pressures or horizontal learning, imitation, and competition mechanisms). However, the review shows that previous literature mainly focuses on the diffusion and development of specific innovations rather than the organisational innovativeness of China's local governments. Theoretically, examining the organisational innovativeness rather than specific innovations could tell us more about the generation of innovations and why some governments are more innovative than others.

On this basis, the book points out that two main structural characteristics of China's bureaucratic organisations could shape local government innovativeness, comprising administrative hierarchy and the span of control. The theoretical analysis first shows that local governments' positions in the administrative hierarchy are positively associated with their innovation resources and negatively associated with their innovation motivations. Since both motivation and resources are necessary but insufficient conditions for generating innovations, combining these two logics, this book proposes an inverse U-shaped relationship between the administrative ladder and local government innovativeness in China. In other words, county- or city-level governments should be more innovative, compared to the provincial- or township-level governments, given their less constrained innovation resources or innovation motivations.

Further theoretical analysis shows that the span of control, proxied by the number of subordinate governments managed by a superior government, can increase local government innovativeness in three ways. First, when the span of control increases, the subordinate governments' policy autonomy increases due to the growing monitoring costs borne by the superior governments, thus increasing the subordinate governments' capability of generating innovations. Second, when the span of control increases, the subordinate governments' political competition intensity increases due to the rising number of competitors and the decreasing opportunities for local leaders to get promoted under the same superior government, thus increasing the subordinate governments' motivations of generating innovations to outperform their peers. Third, when the span of control increases, the subordinate governments' horizontal transfer opportunities within a higher-level jurisdiction also increase, and local leaders have more opportunities to adopt innovative government practices.

Before conducting the empirical analysis, this book provides a detailed review of the local governments' innovative practices in China by examining the Innovations and Excellence in Chinese Local Governance (IECLG) Awards Programme from 2001 to 2015. Inspired by the Innovations in American Government Awards (IAGA) Programme created by Harvard Kennedy School, the IECLG Awards Programme was an independent nonprofit academic programme created in 2000 and first implemented in 2001 by the Central Compilation and Translation Bureau, the Central Party School of the Central Committee of the Chinese Communist Party, and Peking University. This book presents the history of the eight waves of the IECLG Awards Programme and the typology and distribution of the 177 IECLG winners and finalists.

Empirically, this book operationalizes the concept of government innovativeness with the number of IECLG winners and finalists in a jurisdiction. This book analyses the distribution of IECLG winners and finalists across government tiers and years and shows that the majority of those IECLG winners and finalists are city- and county-level governments rather than provincial- or township-level governments, thus providing preliminary empirical support for the theorised relationship between administrative

Preface xi

hierarchy and local government innovativeness. I also created two original panel datasets covering 31 provinces from 2001 to 2015 and 270 cities from 2005 to 2015. A series of provincial- and city-level statistical analyses provide strong evidence for a significantly positive relationship between the number of city-level jurisdictions within a province or the number of county-level jurisdictions within a city—a proxy for the span of control—and the number of city-level IECLG winners and finalists in a province or the number of county-level IECLG winners and finalists in a city.

The theoretical and empirical findings reported in this book have important implications for the development of government innovation theory and managerial practices in the public sector. Unlike previous explanations for local government innovations based on internal determinants or external pressures, this book provides the first attempt to establish theoretical logics that associate the structural characteristics of China's multilevel bureaucratic organisations with local government innovativeness. The exploration of the structural factors underlying local government innovativeness could help us understand why some local governments are more innovative than others. Moreover, given the potentially significant governance performances and socioeconomic consequences created by the variation of local government innovativeness, the exploration of the effects of administrative hierarchy and the span of control on local government innovativeness presented in this book can provide the scholarly foundations for future efforts of examining how institutional contexts shape administrative reforms and socioeconomic development in a country.

I would like to thank my mentors and collaborators, Dr Xufeng Zhu, Dr Manuel P. Teodoro, and Dr Xinsheng Liu for their guidance and support throughout this research. Also, I am grateful to the series editors of Routledge Focus on Public Governance in Asia, Dr Hong Liu from Nanyang Technological University and Dr Wenxuan Yu from Xiamen University, whose support has been important to me in writing this book. I was also extremely fortunate to have had the opportunity to present an earlier version of this book in the virtual speaker series for political science scholars organised by Emerson Niou, Junyan Jiang, Zhaotian Luo, Tianyang Xi, and Qiang Zhou. Thanks also go to my friends, colleagues, and the school faculty and staff for making my time at Renmin University of China a great experience. Finally, thanks to my family for their encouragement, patience, and love.

Funding sources

This research was supported by National Natural Science Foundation of China (Grant Number: 72004220), the Major Project of the National Social

Science Foundation of China (Grant Number: 20ZDA042), the special project of "the interdisciplinary platform for research on governance innovation of megacities with Chinese characteristics in the smart era" at Renmin University of China, and the National Natural Science Foundation of China (Grant Number: 71874198).

Youlang Zhang
Renmin University of China, Beijing, China

Abbreviations

ALC	Administrative Licensing Centre
CAS	Chinese Academy of Science
CCGI	Centre for Chinese Government Innovations at Peking University
CCP	Chinese Communist Party
CCTB	Central Compilation and Translation Bureau
COVID-19	Coronavirus Disease 2019
CPESC	Comparative Politics and Economics Study Centre at the Central Compilation and Translation Bureau
CPS	Central Party School of the Chinese Communist Party
CSCWP	Comparative Study Centre of World Parties at the Central Party School of the Chinese Communist Party
CYLC	Communist Youth League of China
GG	Grid Governance
EHA	Event History Analysis
GDP	Gross Domestic Product
IAGA	Innovations in American Government Awards
IECLG	Innovations and Excellence in Chinese Local Governance
IMA	Innovative Management Award (Canada)
IPAC	Institute of Public Administration of Canada
MPIR	Multiple-Plan Integration Reform
NBS	National Bureau of Statistics
NFA	National Forestry Administration
NMA	National Meteorological Administration
NPM	New Public Management
PA	Public Administration
PAR	Poisson Autoregression Regression
PKU	Peking University
PMC	Province-Managing-County
PP	Public Policy

RCCP	Research Centre for Chinese Politics
REDCP	Resource and Environment Data Cloud Platform
SDR	Super-Department Reform
TPCR	Turning the Prefectures into Cities Reform
UMLSAS	Urban Minimum Living Standard Assistance System

1 Introduction

Government innovations are policies, projects, programmes, or organisational reforms that are new to the government that adopts them (Aiken and Hage 1971; Rogers 2010; Walker, Jeanes, and Rowlands 2002). In general, the adoption of innovation has become an important strategy for governments in order to achieve the goals of improving government performance or regime legitimacy, attracting human and economic capital, and supporting socioeconomic development. For instance, since the 1980s the New Public Management reforms—comprising performance management, output controls, decentralisation, and e-government—have been widely adopted by governments around the world (Hammerschmid et al. 2019; Kaboolian 1998). Moreover, in recent years, various modes of network governance, comprising privatisation, public–private partnerships, and contracting, have become increasingly popular in both developed and developing countries (Bach et al. 2016; Yi et al. 2018).

An increased interest in local government innovation research in China has emerged in recent years (Wu and Zhang 2018; Zhang and Zhu 2020a, 2020b). A key research question for scholars is: Why do some governments adopt an innovation while others do not? To answer this question, studies mainly focused on one specific innovation case and tested the potential effects of jurisdictional attributes, government or leadership characteristics, and intergovernmental interactions on local government innovation adoption (e.g., Ma 2013; Zhang 2012; Zhu 2014; Zhu and Zhang 2016). While the extant research focusing on explaining the adoption of specific innovations has greatly contributed to our understanding of local government decisions and behaviours in China, it tends to ignore an important aspect of local innovation dynamics: the organisational innovativeness of local governments, which refers to the extent to which a government is capable of generating innovations (Ma 2017; Walker, Berry, and Avellaneda 2015). Unlike previous studies—which focused on the adoption of certain specific innovations—this study focuses on explaining government innovativeness and aims to explore the answers to a broader question: Why are some governments more innovative than others? Or, to put it differently, what determines a government's propensity to innovate?

Understanding the logic of government innovativeness is important because it has significant theoretical, normative, and practical implications. On the theoretical side, government innovativeness is directly connected to a core topic in social science: the decision-making of policymakers. Examining government innovativeness could help to produce important theoretical insights into the patterns of how and why some governments systematically adopt more reforms or innovative policy choices than others. More specifically, examining government innovativeness can help us form a better understanding of how governments respond in the long run to the institutional incentives in the political or bureaucratic system, socioeconomic factors in their jurisdictions, peer pressure from other governments, and coercive pressure from their superior governments.

On the normative side, government innovativeness may significantly affect socioeconomic equity. The differing extent of adopting government innovations implies the varying allocations of government and socioeconomic resources, and therefore largely accounts for heterogeneous government performance and political, economic, and social inequity within and across jurisdictions. On the one hand, governments that consistently adopt innovations that favour certain social groups may create enormous vested interest groups that dominate policymaking (Moynihan and Soss 2014; Pierson 1993). On the other hand, governments that consistently adopt innovations to improve their performance may attract more residents and private investment, which is beneficial for their adoption of future innovations, thus creating a positive feedback loop (Pierson 2000). By contrast, governments that fail to adopt innovations to improve their performance may miss the opportunity to obtain the necessary resources for further improvements in government performance and socioeconomic development.

Finally, on the practical side, examining government innovativeness could generate important policy lessons for practitioners. As previous literature shows, some governments tend to have an unfair advantage over other governments regarding innovations, such as the State of California in the United States or the Zhejiang province in China (Volden 2006; Zhao 2012). Systematic theoretical and empirical analyses of the determinants of government innovativeness can help reformers to identify the institutional obstacles to innovations, predict the demand for innovations, manage innovation activities, and promote the diffusion of the best institutional or organisational designs in public administration and public policy across governments.

The major goal of this study was to provide an institutional explanation for local government innovativeness in China. In particular, this study focused on the variation of the span of control across China (e.g., the number of city-level jurisdictions within a province or the number of county-level jurisdictions within a city) and developed a novel theoretical framework that explains the effect of the structural characteristics of China's bureaucratic organisations on local government innovativeness and the underlying mechanisms. This study points out that in China's multi-level government structure, the administrative hierarchy and the span of control could shape local governments' innovation motivations, innovation capabilities, and innovation opportunities, thus systematically determining local government innovativeness.

Theoretically, this book contributes to public administration (PA), public policy (PP), and political science literature by offering theoretical and empirical insights into the structural factors that influence local government innovativeness. Previous explanations for local government innovation choices or behaviours mainly focus on contextual characteristics, leader characteristics, or external diffusional pressures (Zhang and Zhu 2020b). The administrative hierarchy and the span of control, as classic topics in management science (Gulick 1937; Meier and Bohte 2000; Simon 1997), are important but often overlooked structural elements of China's multi-level government system in local government innovation studies. However, previous research suggests that the administrative hierarchy and the span of control can significantly affect officials' career incentives, organisational stability, and governance complexity (Benson 1977; Cathcart et al. 2004; Keren and Levhari,1979; Landry, Lü, and Duan 2018; Lü and Landry 2014; Sahay and Gupta 2011). Thus, it is reasonable to expect these structural characteristics to affect local innovative behaviours too. Examining local government innovativeness through the lens of the administrative hierarchy and the span of control can help to develop new theories to improve the academic understanding of local government decisions and behaviours.

Empirically, this book offers new evidence on local government innovativeness in China. Based on eight waves of the biennial Innovations and Excellence in Chinese Local Governance (IECLG) Awards Programme from 2001 to 2015, I present two newly created original panel datasets covering 31 provinces from 2001 to 2015 and 270 cities from 2005 to 2015 by collecting archived statistical information from multiple sources and manual coding. I then provide detailed descriptions of the operation, distributions, and trends of the local government innovations in China. In addition to the descriptive analysis as conducted by previous research on IECLG (Wu, Ma, and Yang 2013), this book also provides a detailed large-N statistical analysis to identify the determinants of local government innovativeness.

4 Introduction

Practically, the analyses and findings of the book can provide important guidance for practitioners regarding administrative reforms and improving government innovativeness. In recent years, China has been promoting administrative division reforms (e.g., the "Province-Managing-County Reform" or the "Turning the Prefectures into Cities Reform") and departmental reforms (e.g., the "Super-Department Reform" or "Multiple-Plan Integration Reform") that are directly relevant to changes in the administrative hierarchy or the span of control (Chung and Lam 2004; Liu et al. 2018; Ma 2005; Ma 2016; Yew 2012). This book will help practitioners to predict potential consequences of these reforms in terms of the frequency or consistency of local governments' innovative choices and behaviours.

This rest of the book consists of five chapters structured to answer the key research question: Why are some governments more innovative than others? The second chapter provides a detailed review of previous empirical studies on local government innovation adoption and local government innovativeness in China, paying particular attention to the multiple categories of explanatory variables. The third chapter then develops a theoretical framework for analysing how the structural characteristics of bureaucratic organisations in China affect local government innovativeness. The fourth chapter provides an overview of eight waves of the IECLG Awards Programme from 2001 to 2015. The fifth chapter provides a multi-level empirical analysis to evaluate the aforementioned theoretical framework. The sixth chapter discusses the theoretical and practical implications of the main findings and lays out an agenda for future research.

Bibliography

Aiken, Michael, and Jerald Hage. "The organic organization and innovation." *Sociology* 5, no. 1 (1971): 63–82.

Bach, Tobias, Fabrizio De Francesco, Martino Maggetti, and Eva Ruffing. "Transnational bureaucratic politics: An institutional rivalry perspective on EU network governance." *Public Administration* 94, no. 1 (2016): 9–24.

Benson, J. Kenneth. "Innovation and crisis in organizational analysis." *The Sociological Quarterly* 18, no. 1 (1977): 3–16.

Cathcart, Deb, Susan Jeska, Joan Karnas, Sue E. Miller, Judy Pechacek, and Lolita Rheault. "Span of control matters." *JONA: The Journal of Nursing Administration* 34, no. 9 (2004): 395–399.

Chung, Jae Ho, and Tao-chiu Lam. "China's "city system" in flux: Explaining postmao administrative changes." *The China Quarterly* 180 (2004): 945–964.

Gulick, Luther. "Notes on the theory of organization." *Classics of Organization Theory* 3, no. 1937 (1937): 87–95.

Hammerschmid, Gerhard, Steven Van de Walle, Rhys Andrews, and Ahmed Mohammed Sayed Mostafa. "New public management reforms in Europe and their effects: Findings from a 20-country top executive survey." *International Review of Administrative Sciences* 85, no. 3 (2019): 399–418.

Kaboolian, Linda. "The new public management: Challenging the boundaries of the management vs. administration debate." *Public Administration Review* 58, no. 3 (1998): 189–193.

Keren, Michael, and David Levhari. "The optimum span of control in a pure hierarchy." *Management Science* 25, no. 11 (1979): 1162–1172.

Landry, Pierre F., Xiaobo Lü, and Haiyan Duan. "Does performance matter? Evaluating political selection along the Chinese administrative ladder." *Comparative Political Studies* 51, no. 8 (2018): 1074–1105.

Liu, Yanxu, Bojie Fu, Wenwu Zhao, Shuai Wang, and Yu Deng. "A solution to the conflicts of multiple planning boundaries: Landscape functional zoning in a resource-based city in China." *Habitat International* 77 (2018): 43–55.

Lü, Xiaobo, and Pierre F. Landry. "Show me the money: Interjurisdiction political competition and fiscal extraction in China." *American Political Science Review* 108, no. 3 (2014): 706–722.

Ma, Laurence J. C. "Urban administrative restructuring, changing scale relations and local economic development in China." *Political Geography* 24, no. 4 (2005): 477–497.

Ma, Liang. "The diffusion of government microblogging: Evidence from Chinese municipal police bureaus." *Public Management Review* 15 (2013): 288–309.

Ma, Liang. "Does super-department reform improve public service performance in China?" *Public Management Review* 18, no. 3 (2016): 369–391.

Ma, Liang. "Political ideology, social capital, and government innovativeness: Evidence from the US states." *Public Management Review* 19, no. 2 (2017): 114–133.

Meier, Kenneth J., and John Bohte. "Ode to Luther Gulick: Span of control and organizational performance." *Administration & Society* 32, no. 2 (2000): 115–137.

Moynihan, Donald P., and Joe Soss. "Policy feedback and the politics of administration." *Public Administration Review* 74, no. 3 (2014): 320–332.

Pierson, Paul. "When effect becomes cause: Policy feedback and political change." *World Politics* 45, no. 4 (1993): 595–628.

Pierson, Paul. "Increasing returns, path dependence, and the study of politics." *American Political Science Review* 91, no. 2 (2000): 251–267.

Rogers, Everett M. *Diffusion of innovations*. New York: Simon and Schuster, 2010.

Sahay, Yamini Prakash, and Meenakshi Gupta. "Role of organization structure in innovation in the bulk-drug industry." *Indian Journal of Industrial Relations* 46, no. 3 (2011): 450–464.

Simon, Hebert. *Administrative behavior: A study of decision-making processes in administrative organizations*. New York: Free Press, 1997.

Volden, Craig. "States as policy laboratories: Emulating success in the children's health insurance program." *American Journal of Political Science* 50, no. 2 (2006): 294–312.

Walker, Richard M., Frances S. Berry, and Claudia N. Avellaneda. "Limits on innovativeness in local government: Examining capacity, complexity, and dynamism in organizational task environments." *Public Administration* 93, no. 3 (2015): 663–683.

Walker, Richard M., Emma Jeanes, and Robert Rowlands. "Measuring innovation–applying the literature-based innovation output indicator to public services." *Public Administration* 80, no. 1 (2002): 201–214.

Wu, Jiannan, Liang Ma, and Yuqian Yang. "Innovation in the Chinese public sector: Typology and distribution." *Public Administration* 91, no. 2 (2013): 347–365.

Wu, Jiannan, and Pan Zhang. "Local government innovation diffusion in China: An event history analysis of a performance-based reform programme." *International Review of Administrative Sciences* 84, no. 1 (2018): 63–81.

Yew, Chiew Ping. "Pseudo-urbanization? Competitive government behavior and urban sprawl in China." *Journal of Contemporary China* 21, no. 74 (2012): 281–298.

Yi, Hongtao, Frances Stokes Berry, and Wenna Chen. "Management innovation and policy diffusion through leadership transfer networks: An agent network diffusion model." *Journal of Public Administration Research and Theory* 28, no. 4 (2018): 457–474.

Zhang, Yanlong. Institutional sources of reform: The diffusion of land banking systems in China. *Management and Organization Review* 8 (2012): 507–33.

Zhang, Youlang, and Xufeng Zhu. "The moderating role of top-down supports in horizontal innovation diffusion." *Public Administration Review* 80, no. 2 (2020a): 209–221.

Zhang, Youlang, and Xufeng Zhu. "Career cohorts and inter-jurisdictional innovation diffusion: An empirical exploration in China." *International Public Management Journal* 23, no. 3 (2020b): 421–441.

Zhao, Qiang. "The regional disparities in Chinese provincial government innovation." *Innovation* 14, no. 4 (2012): 595–604.

Zhu, Xufeng. "Mandate versus championship: Vertical government intervention and diffusion of innovation in public services in authoritarian China." *Public Management Review* 16 (2014): 117–39.

Zhu, Xufeng, and Youlang Zhang. "Political mobility and dynamic diffusion of innovation: The spread of municipal pro-business administrative reform in China." *Journal of Public Administration Research and Theory* 26, no. 3 (2016): 535–551.

2 Explanations for local government innovation in China

To situate this study in relation to existing knowledge, this chapter provides a detailed overview of previous empirical findings on China's local government innovative choices and behaviours published in English to help the readers understand what has and has not been learnt about these topics. This chapter pays particular attention to studies on local government innovation adoption and local government innovativeness to lay the foundation and provide context for the theoretical and empirical analyses in later chapters.

As Figure 2.1 shows, since the 1960s, extensive studies in the disciplines of economics, management science, political science, sociology, PA, and PP have examined government innovations (Damanpour 1991; Mohr 1969; Walker 1969; Wolfe 1994). Most early literature focuses on cases in developed countries and, particularly, the United States (Graham, Shipan, and Volden 2013). Scholars have widely discussed the effects of jurisdictional characteristics, government characteristics, innovation attributes, and various external pressures on government innovation (Boushey 2010; Shipan and Volden 2012). Notably, another concept similar to government innovation is policy innovation. Although these two concepts have almost the same meaning, the concept of government innovation is mainly used by scholars who focus on public management and draw insights from organisational theories (Damanpour 1991; Wolfe 1994), while the concept of policy innovation is mainly used by scholars who focus on politics and policy process and draw insights from policy diffusion theory (Walker 1969; Berry and Berry 1990). However, in recent years, these two streams of literature have been increasingly mixed and combined (Walker, Berry, and Avellaneda 2015). To fully review the existing literature on governments' innovative practices, this chapter examines published empirical studies emphasising either government innovation or policy innovation. Moreover, to simplify, this book uses the concept of government innovation to represent either government innovation or policy innovation.

8 *Explanations for local innovation*

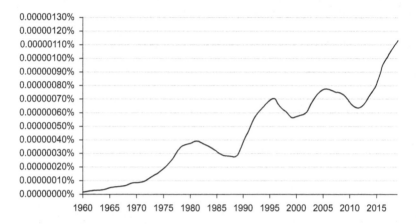

Figure 2.1 The frequency of "government innovation" found in sources printed between 1960 and 2019 in Google's text corpora (Google Books Ngram Viewer)

Over the past two decades, some scholars have begun to focus their attention on government innovations in China. China has a context that radically differs from the Western industrialised democracies regarding socioeconomic development, political institutions, and cultural traditions (Zhang 2019; Zhang and Zhu 2019). Hence, the existing theoretical and empirical findings cannot be indistinguishably applied to explain China's government innovations (Berry and Berry 1999; Walker, Avellaneda, and Berry 2011; Zhu 2014). In recent years, scholars have provided detailed descriptions and categorisations of China's government innovations by examining the winners of the IECLG Awards Programme, initiated and organised by the Central Compilation and Translation Bureau (CCTB), Central Party School of the Chinese Communist Party (CPSCCP), and Peking University (Chen and Göbel 2016; Wu, Ma, and Yang 2013). Many scholars believe that the widespread adoption of government innovations in China has been an important driver of China's economic success and social stability in the past decades (Heilmann 2008a, 2008b; Wang 2009).

Nevertheless, it was not until the 2010s that scholars attempted to use empirical data systematically to answer the following question: How and why do local government innovations occur in China? As summarised in Figure 2.2, most of the previous studies focused on innovation adoption and extant explanations can be classified into two main categories: internal determinants and external pressures. The internal determinants include macro-level jurisdictional and government characteristics and micro-level

Explanations for local innovation 9

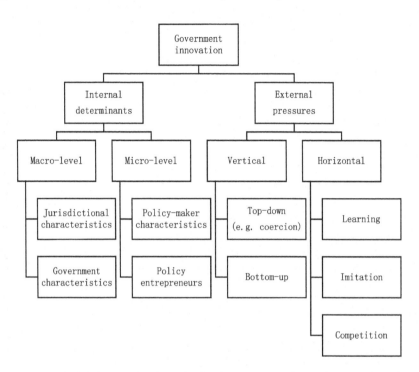

Figure 2.2 Existing explanations for local government innovation adoption in China

policymakers' or policy entrepreneurs' characteristics. For instance, Ma (2013, 2014) found that government size, Internet penetration rate, fiscal resources, and information technology (IT) capacity are positively associated with the adoption and assimilation of government microblogging. Zhu and Zhang (2016) showed that the career mobility of city leaders significantly influences their decisions about creating administrative licensing centres (ALC). Wu and Zhang (2018) showed that provincial leaders' relative age and chances of promotion are significantly negatively associated with the adoption of performance-based reform programmes. Based on in-depth qualitative analysis, Zhu Y. (2012) and Zhu X. (2018) highlighted the central role of policy entrepreneurship in fostering local government innovations.

Multiple types of external pressures could affect government innovation adoption or diffusion too. The potential vertical external pressures include the top-down effect and the bottom-up effect. For instance, Zhu and Zhang (2019) showed that the top-down intervention policies from the central and

10 Explanations for local innovation

provincial governments could stimulate the spread of marketisation innovation among Chinese cities. Nevertheless, Zhang and Zhu (2019) found no consistently significant evidence to support the bottom-up effect. Potential horizontal external pressures include the learning effect, imitation effect, and competition effect. Multiple studies have shown that various types of peer pressure could significantly promote the diffusion of government innovations (Zhang 2012, 2015; Zhang and Zhu 2020a, 2020b; Zhu and Zhao 2018).

Empirically, most studies have focused on the reform of public service institutions and agencies. Specifically, studies have examined the land banking system (Zhang 2012), housing monetisation reform (Zhu 2012), government microblogging (Ma 2013, 2014), administrative licensing reform (Zhu and Zhang 2016, 2019; Zhang and Zhu 2019), public–private partnership (Zhang 2015), performance management reform (Liu and Li 2016; Wu and Zhang 2018), the Urban Minimum Living Standard Assistance System (UMLSAS) (Zhu and Zhao 2018), environmental regulation (Zhang and Wu 2020), and pension policies (Zhu and Zhao Early view).

Methodologically, empirical studies have employed either quantitative or qualitative methods to generate evidence for their arguments. The quantitative studies mainly used event history analyses (EHA) with the semiparametric Cox proportional hazards model (Zhang 2012), the parametric logit model or probit model (Wu and Zhang 2018; Zhu and Zhang 2016, 2019), multi-level mixed-effects model (Zhang 2015), and spatial autoregression models (Yi, Berry, and Chen 2018; Zhang and Wu 2020). Researchers have employed these quantitative methods to examine cross-sectional (Ma 2013, 2014) and time-series cross-sectional datasets (Zhang and Zhu 2020a, 2020b). Qualitative studies have used single-case studies (Zhu 2012, 2013), within-case comparative analysss (Zhu 2018), and comparative case studies designs (Zhu 2014, 2017; Zhu and Bai 2020; Zhu and Zhao Early view). In general, both quantitative and qualitative studies have widely examined the national-, provincial-, and city-level adoption of specific government innovations.

Table 2.1 provides an overview of some explanatory empirical studies on local government innovation adoption in China, listed by publication year. As Table 2.1 shows, the explanatory empirical studies on local government innovation in China have been frequently published in the mainstream disciplinary journals of PA and PP in the last ten years. This finding suggests that the theoretical or empirical insights drawn from studies on local government innovation in China have been increasingly recognised and valued by the international academic community.

Despite an increasing scholarly interest in examining innovation adoption in the public sector in China (Zhu and Zhang 2020a), only one study

Table 2.1 An overview of some explanatory empirical studies published in the mainstream disciplinary journals of PA and PP (listed by publication year)

Paper	Journal	Policy case	Empirical methods	Main findings (excerpted from the abstract)
Yanlong Zhang (2012)	Management and Organization Review	City-level land banking system	EHA with Cox models	"Local economic characteristics have a declining effect on the adoption of land banking systems and gradually yield to pressure from peer cities, provincial governments, policy professionals, and policy-making communities."
Yapeng Zhu (2012)	Australian Journal of Public Administration	Provincial-level housing monetarisation reform	Single case study	"The key decision-maker, being a policy entrepreneur, plays a central role in fostering policy innovations. An important aspect of the policy entrepreneur's strategy is to promote civic engagement and to strengthen legitimacy by leveraging the popularity of innovative policy."
Liang Ma (2013)	Public Management Review	City-level police microblogging	Cross-sectional analysis with logit and Tobit models	"Government size, internet penetration rate, regional diffusion effects and upper-tier pressure are positively and significantly associated with the adoption and earliness of police microblogging, whereas fiscal revenue, economic development and openness, E-government and public safety have no significant effects."
Liang Ma (2014)	Public Management Review	City-level government microblogging	Cross-sectional analysis with negative binomial models	"Horizontal competition is found to be significantly and positively associated with the assimilation of government microblogging … the results support the significantly positive effects of fiscal resources and IT capacity. Municipal wealth, size and administrative ranking are also positively and significantly correlated with the number of government microblogs."

(*Continued*)

Table 2.1 Continued

Paper	Journal	Policy case	Empirical methods	Main findings (excerpted from the abstract)
Xufeng Zhu (2014)	Public Management Review	City-level administrative licensing service reforms	Comparative case study	"Administrative commands facilitate the formation of the 'mandatory policy diffusion' that rapidly diffuses policy instruments. Competition in the performance evaluation-based personnel system contributes to the formation of 'championship policy diffusion', which leads to the divergence of policy instruments in neighbouring local governments."
Yanlong Zhang (2015)	Journal of Public Policy	City-level public-private partnership	Three-level multilevel mixed-effects models	"The spatial effects appear to be significantly modulated when the influence from structurally equivalent peer cities are considered; moreover, the effects of the vertical diffusion mechanisms are moderated by the liberalisation index of the contract forms, and the horizontal diffusion mechanisms are moderated by the marketability of the infrastructure segments."
Xufeng Zhu and Youlang Zhang (2016)	Journal of Public Administration Research and Theory	City-level administrative licensing centres	EHA with logit models and piecewise constant exponential models	"Without central authority intervention, innovation adoption tended to be similar to an "economic decision" based on city characteristics. After the accession of China to the World Trade Organization, the mechanism of "neighboring diffusion" dominated the innovation adoption process. Otherwise, after the central authority promulgated the Administrative Permission Law in 2004, innovation adoption became similar to a "political decision" made by local officials concerned about their personal political mobility."
Wei Liu and Wenzhao Li (2016)	Public Performance & Management Review	City-level performance management campaign	Comparative case study	"When the contents of diffusion are different, the superior government applies different levels of coercive power, and the mechanisms, characteristics, and outcomes of diffusion also differ."

Jiannan Wu and Pan Zhang (2018)	*International Review of Administrative Sciences*	Provincial-level performance-based reform programme	EHA with probit models	"Leaders' relative age and chances of being appointed to the Politburo, and distance to the general election, are significantly negatively correlated with the reform programme's adoption, but top-down diffusion is significantly positively correlated with it."
Hongtao Yi, Frances Stokes Berry, and Wenna Chen (2018)	*Journal of Public Administration Research and Theory*	Provincial-level energy performance	Spatial autoregression models	"Leadership transfer networks channel performance innovation between locations where managers served/serve, especially when the institutional environments are similar between the locations."
Xufeng Zhu (2018)	*Journal of Comparative Policy Analysis: Research and Practice*	Government reforms within the New Public Management (NPM) paradigm	Within-case comparative analysis	"China's distinctive, geographic approach to vertical and horizontal personnel mobility motivates executive entrepreneurs and energizes transferable learning of specific policy goals and instruments as well as broader paradigms of policy concepts, ideas, principles, and ideologies."
Xufeng Zhu and Hui Zhao (2018)	*Governance*	City-level Urban Minimum Living Standard Assistance system	EHA with logit models and piecewise constant exponential models	"During the era of fiscal recentralization starting from the Chinese Tax-Sharing System Reform in 1994, cities with higher fiscal dependency are more likely to behave innovatively by adopting a new welfare policy for potential fiscal transfer rewards. The central government's recognition of this innovation stimulates cities' adoption but would reverse the effects of fiscal dependency because of the loss of the "innovativeness" of the adoption and its effectiveness in attracting the attention of superior authorities."

(*Continued*)

Table 2.1 Continued

Paper	Journal	Policy case	Empirical methods	Main findings (excerpted from the abstract)
Xufeng Zhu and Youlang Zhang (2019)	Journal of Public Administration Research and Theory	City-level administrative licensing centres	EHA with logit models	"Although the intervention policies from the central or provincial governments independently stimulate the city adoption of marketization innovation, their combined impact on city governments tend to be competitive rather than complementary."
Youlang Zhang and Xufeng Zhu (2019)	Public Management Review	Provincial-level administrative licensing centres	Directed dyadic EHA with logit models	"Horizontal learning, imitation, and the vertical top-down diffusion mechanisms can coexist in China, which provides substantial empirical support for the application of policy diffusion theory in nonwestern countries."
Ciqi Mei and Xiaonan Wang (2020)	Journal of Comparative Policy Analysis: Research and Practice	National-level promotion of policy innovations	Poisson Autoregression Regression (PAR)	"The central authority in China tends to promote fewer sensitive political experiments when inflation increases and resumes promoting experiments when the inflation rate passes a certain tipping point. It is also found that the central authority intentionally regulates the promotion of political experiments during important political events."
Pan Zhang and Jiannan Wu (2020)	International Public Management Journal	Provincial-level atmospheric pollutant emission standards	Spatial autoregression models	"Top-down performance targets can drive the adoption of atmospheric pollutant emission standards in Chinese provinces. Furthermore, the influence of vertical environmental performance targets on the adoption of local atmospheric pollutant emission standards is stronger in Chinese provinces that have adopted more similar standards in the previous period."
Youlang Zhang and Xufeng Zhu (2020a)	Public Administration Review	City-level administrative licensing centres	EHA with logit models	"Results affirm that the effects of different types of diffusional pressures can be conditional on one another—that is, increases in top-down policy supports may substitute the effects of horizontal pressures."

Youlang Zhang and Xufeng Zhu (2020b)	*International Public Management Journal*	Provincial-level administrative licensing centres	Directed dyadic EHA with logit models	"After controlling for multiple previously identified diffusion mechanisms and internal determinants, the cohort effect among provincial party chiefs significantly increases the likelihood of innovation diffusion in China."
Conghu Wang, Xiaoming Li, Wenjuan Ma, and Xiaopeng Wang (2020)	Public Administration and Development	Provincial-level and city-level public resources trading platforms	Comparative case study	"Our analysis and results show that the diffusion models evolve over the different stages of a life cycle of an innovation, contrasting to the literature results that diffusion models remain the same for their studied innovations … We find a first bottom-up and then top-down synthesis approach as an effective, efficient diffusion process for both fitting local needs (i.e. effective) and adopting innovations rapidly nationwide (i.e. efficient)."
Xufeng Zhu and Hui Zhao (Early view)	*Policy Studies Journal*	National-level pension policies	Comparative case study	"First, policy goals and instruments are formed separately and interactively by the central and local governments. Second, the central government is burdened with its own concerns about policy performance for maintaining authority and legitimacy. Third, the evaluation of policy pilots relies primarily on the responses of local governments."

16 *Explanations for local innovation*

has empirically explored the determinants of government innovativeness in China. Using the dataset of the first six waves of the biennial IECLG Awards Programme from 2001 to 2012, Zhao (2012) found that regional economies, degree of openness, fiscal resources, and administrative expenditures are positively associated with the number of winners and finalists in each province. However, Zhao's (2012) analysis might suffer from obvious and serious omitted variable biases as his model specification does not include control variables. Moreover, as Ma (2017) pointed out, Zhao's (2012) indicator of provincial government innovativeness includes innovations initiated by multiple tiers of government, which may introduce serious measurement errors. Further, Zhao (2012) did not take the potential effects of institutional characteristics into consideration.

These studies provide an important scholarly basis for us to understand the history or operations of various local government innovation activities in China. However, the literature generally suffers from three major limitations. First, it mainly focuses on the diffusion and development rather than the generation of government innovations in China (Song et al. 2020). This is a serious omission, because much of the argument of innovation diffusion is based on the premise that the government innovations have already been generated elsewhere. Moreover, even if there was evidence of innovation diffusion, we would like to further identify the determinants of the generation of government innovations in the first place. This begs the following questions: Why are some governments more likely to generate innovations that others? Which factors systematically determine government innovativeness?

Theoretically, examining organisational innovativeness, rather than specific innovations, could tell us more about the generation of innovations. Some scholars have explored the potential determinants of government innovativeness based on cases in Western democracies. For instance, based on the case of the Innovative Management Award (IMA) of the Institute of Public Administration of Canada (IPAC), Bernier, Hafsi, and Deschamps (2015) found that the economy, size of the civil service, deficits, unemployment rate, and type of government could be important determinants of government innovativeness. Using survey information from the corporate and service officers of English local government, Walker, Berry, and Avellaneda (2015) pointed out that the capacity, complexity, and dynamism in organisational task environments may have an impact on managers' perceived organisational innovativeness. Based on a dataset of the finalists and winners recognised by the Innovations in American Government Awards (IAGA) programme, Ma (2017) suggested that government ideology, citizen ideology, and social capital have a positive effect on government innovativeness, measured by the number of state-level award-winning innovations. Nevertheless, although scholars have tried to explain government innovativeness based on the cases in Western democracies, the theoretical and empirical findings cannot be directly applied to China without considering its unique contextual characteristics.

Second, there has been little research into the effect of the structural characteristics of bureaucratic organisations on government innovativeness. The multi-level governance system in modern countries can be characterised by two structural characteristics: the administrative hierarchy and the span of control. The administrative hierarchy refers to the tiers of a government in a multi-level government system. Intuitively, government leaders along the administrative hierarchy may have significantly differing resources, capabilities, and motivations (Kung and Chen 2011; Ma 2005; Xu 2011; Yao and Zhang 2015). For instance, Landry, Lü, and Duan (2018) show that economic performance is more important for bureaucratic promotion at lower levels of government than at higher ones in China. Therefore, it is plausible that the administrative hierarchy could significantly shape government innovativeness.

The span of control refers to the breadth of subordinates managed by a superior (Gulick 1937; Ouchi and Dowling 1974). More specifically, in China's multi-level hierarchical governance system, the span of control could be understood as the number of subordinate governments managed by a superior government (Lü and Landry 2014). For instance, if province A supervises 25 cities while province B supervises 15 cities, then province A has a broader span of control than province B. The span of control is a classic topic in management science and is often connected with organisational innovation and performance, given its potential effects on organisational structure, organisational size, managerial communication, organisational resource allocation, career mobility, and subordinates' autonomy and participation willingness (Benson 1977; Cathcart et al. 2004; Gulick 1937; Keren and Levhari 1979; Sahay and Gupta 2011). In recent years, several scholars have also pointed out that the span of control might affect local governments peer competition, functional responsibility, and fiscal extraction (Lü and Landry 2014). Therefore, it is reasonable to expect that span of control can significantly affect local government innovativeness.

However, to our knowledge, not much research has systematically analysed the association between the administrative hierarchy or the span of control and local government innovativeness in China. The extant research has tended to focus on the internal determinants and external pressures, and assumed that the administrative hierarchy or the span of control would not affect a government's innovation choices. In this study, by contrast, I expected that examining the effect of these structural characteristics of a multi-level governance system on local government innovativeness could help researchers and practitioners to form a deeper understanding of the structural incentives and constraints underlying government choices and behaviours.

Third, most empirical studies have conducted qualitative or quantitative analyses by focusing on a limited number of innovation cases. For instance, all quantitative studies listed in Table 2.1 focused only on the adoption and diffusion of one innovation case. The qualitative studies, such as Zhu (2014), Liu and Li (2016), or Zhu and Zhao (Early view), chose to examine several cases of policy tools in the same policy field. However, only studying one or several innovation cases in a study limits the analysis of organisational-level innovativeness. Moreover, this situation also limits the comparability of empirical findings based on different cases drawn from different research articles.

Moreover, apart from Wu, Ma, and Yang (2013), research has provided few updated descriptions of the trend of the innovative practices of local governments in China. Wu, Ma, and Yang's (2013) analysis only examines 83 winners and finalists in the IECLG Awards Programme between 2001 and 2008. However, there were 94 new winners and finalists in the IECLG Awards Programme between 2009 and 2016. Therefore, in addition to providing a new explanatory framework for local government innovativeness in China, another important goal of this study was to provide an updated analysis of the general trend of the innovative practices of local governments in China based on all winners and finalists in the IECLG Awards Programme between 2001 and 2016.

Bibliography

Benson, J. Kenneth. "Innovation and crisis in organizational analysis." *The Sociological Quarterly* 18, no. 1 (1977): 3–16.

Bernier, Luc, Taïeb Hafsi, and Carl Deschamps. "Environmental determinants of public sector innovation: A study of innovation awards in Canada." *Public Management Review* 17, no. 6 (2015): 834–856.

Berry, Frances Stokes, and William D. Berry. "State lottery adoptions as policy innovations: An event history analysis." *American Political Science Review* 84, no. 2 (1990): 395–415.

Berry, Frances Stokes, and William D. Berry. "Innovation and diffusion models in policy research." In P.A. Sabatier (ed.), *Theories of the Policy Process*, Boulder, CO: Westview, 1999.

Boushey, Graeme. *Policy diffusion dynamics in America*. New York: Cambridge University Press, 2010.

Cathcart, Deb, Susan Jeska, Joan Karnas, Sue E. Miller, Judy Pechacek, and Lolita Rheault. "Span of control matters." *JONA: The Journal of Nursing Administration* 34, no. 9 (2004): 395–399.

Chen, Xuelian, and Christian Göbel. "Regulations against revolution: Mapping policy innovations in China." *Journal of Chinese Governance* 1, no. 1 (2016): 78–98.

Damanpour, Fariborz. "Organizational innovation: A meta-analysis of effects of determinants and moderators." *Academy of Management Journal* 34, no. 3 (1991): 555–590.

Graham, Erin R., Charles R. Shipan, and Craig Volden. "The diffusion of policy diffusion research in political science." *British Journal of Political Science* 43, no. 3 (2013): 673–701.

Gulick, Luther. "Notes on the theory of organization." *Classics of Organization Theory* 3, no. 1937 (1937): 87–95.

Heilmann, Sebastian. "Policy experimentation in China's economic rise." *Studies in Comparative International Development* 43, no. 1 (2008a): 1–26.

Heilmann, Sebastian. "From local experiments to national policy: The origins of China's distinctive policy process." *The China Journal* 59 (2008b): 1–30.

Keren, Michael, and David Levhari. "The optimum span of control in a pure hierarchy." *Management Science* 25, no. 11 (1979): 1162–1172.

Kung, James Kai-Sing, and Shuo Chen. "The tragedy of the nomenklatura: Career incentives and political radicalism during China's Great Leap famine." *American Political Science Review* 105, no. 1 (2011): 27–45.

Landry, Pierre F., Xiaobo Lü, and Haiyan Duan. "Does performance matter? Evaluating political selection along the Chinese administrative ladder." *Comparative Political Studies* 51, no. 8 (2018): 1074–1105.

Liu, Wei, and Wenzhao Li. "Divergence and convergence in the diffusion of performance management in China." *Public Performance & Management Review* 39, no. 3 (2016): 630–654.

Lü, Xiaobo, and Pierre F. Landry. "Show me the money: Interjurisdiction political competition and fiscal extraction in China." *American Political Science Review* 108, no. 3 (2014): 706–722.

Ma, Laurence JC. "Urban administrative restructuring, changing scale relations and local economic development in China." *Political Geography* 24, no. 4 (2005): 477–497.

Ma, Liang. "The diffusion of government microblogging: Evidence from Chinese municipal police bureaus." *Public Management Review* 15, no. 2 (2013): 288–309.

Ma, Liang. "Diffusion and assimilation of government microblogging: Evidence from Chinese cities." *Public Management Review* 16, no. 2 (2014): 274–295.

Ma, Liang. "Political ideology, social capital, and government innovativeness: Evidence from the US states." *Public Management Review* 19, no. 2 (2017): 114–133.

Mei, Ciqi, and Xiaonan Wang. "Wire-walking: Risk management and policy experiments in China from a comparative perspective." *Journal of Comparative Policy Analysis: Research and Practice* 22, no. 4 (2020): 360–382.

Mohr, Lawrence B. "Determinants of innovation in organizations." *American Political Science Review* 63, no. 1 (1969): 111–126.

Ouchi, William G., and John B. Dowling. "Defining the span of control." *Administrative Science Quarterly* 19, no. 3 (1974): 357–365.

Sahay, Yamini Prakash, and Meenakshi Gupta. "Role of organization structure in innovation in the bulk-drug industry." *Indian Journal of Industrial Relations* 46, no. 3 (2011): 450–464.

Shipan, Charles R., and Craig Volden. "Policy diffusion: Seven lessons for scholars and practitioners." *Public Administration Review* 72, no. 6 (2012): 788–796.

Song, Qijiao, Ming Qin, Ruichen Wang, and Ye Qi. "How does the nested structure affect policy innovation?: Empirical research on China's low carbon pilot cities." *Energy Policy* 144 (2020): 111695.

Walker, Jack L. "The diffusion of innovations among the American states." *American Political Science Review* 63, no. 3 (1969): 880–899.

Walker, Richard M., Claudia N. Avellaneda, and Frances S. Berry. "Exploring the diffusion of innovation among high and low innovative localities: A test of the Berry and Berry model." *Public Management Review* 13, no. 1 (2011): 95–125.

Walker, Richard M., Frances S. Berry, and Claudia N. Avellaneda. "Limits on innovativeness in local government: Examining capacity, complexity, and dynamism in organizational task environments." *Public Administration* 93, no. 3 (2015): 663–683.

Wang, Conghu, Xiaoming Li, Wenjuan Ma, and Xiaopeng Wang. "Diffusion models over the life cycle of an innovation: A bottom-up and top-down synthesis approach." *Public Administration and Development* 40, no. 2 (2020): 105–118.

Wang, Shaoguang. "Adapting by learning: The evolution of China's rural health care financing." *Modern China* 35, no. 4 (2009): 370–404.

Wolfe, Richard A. "Organizational innovation: Review, critique and suggested research directions." *Journal of Management Studies* 31, no. 3 (1994): 405–431.

Wu, Jiannan, Liang Ma, and Yuqian Yang. "Innovation in the Chinese public sector: Typology and distribution." *Public Administration* 91, no. 2 (2013): 347–365.

Wu, Jiannan, and Pan Zhang. "Local government innovation diffusion in China: An event history analysis of a performance-based reform programme." *International Review of Administrative Sciences* 84, no. 1 (2018): 63–81.

Xu, Chenggang. "The fundamental institutions of China's reforms and development." *Journal of Economic Literature* 49, no. 4 (2011): 1076–1151.

Yao, Yang, and Muyang Zhang. "Subnational leaders and economic growth: Evidence from Chinese cities." *Journal of Economic Growth* 20, no. 4 (2015): 405–436.

Yi, Hongtao, Frances Stokes Berry, and Wenna Chen. "Management innovation and policy diffusion through leadership transfer networks: An agent network diffusion model." *Journal of Public Administration Research and Theory* 28, no. 4 (2018): 457–474.

Zhang, Pan, and Jiannan Wu. "Performance targets, path dependence, and policy adoption: Evidence from the adoption of pollutant emission control policies in Chinese provinces." *International Public Management Journal* 23, no. 3 (2020): 405–420.

Zhang, Yanlong. "Institutional sources of reform: The diffusion of land banking systems in China." *Management and Organization Review* 8 (2012): 507–533.

Zhang, Yanlong. "The formation of public-private partnerships in China: An institutional perspective." *Journal of Public Policy* 35, no. 2 (2015): 329.

Zhang, Youlang. "Representative bureaucracy, gender congruence, and student performance in China." *International Public Management Journal* 22, no. 2 (2019): 321–342.

Zhang, Youlang, and Xufeng Zhu. "Multiple mechanisms of policy diffusion in China." *Public Management Review* 21, no. 4 (2019): 495–514.

Zhang, Youlang, and Xufeng Zhu. "The moderating role of top-down supports in horizontal innovation diffusion." *Public Administration Review* 80, no. 2 (2020a): 209–221.

Zhang, Youlang, and Xufeng Zhu. "Career cohorts and inter-jurisdictional innovation diffusion: An empirical exploration in China." *International Public Management Journal* 23, no. 3 (2020b): 421–441.

Zhao, Qiang. "The regional disparities in Chinese provincial government innovation." *Innovation* 14, no. 4 (2012): 595–604.

Zhu, Xufeng. "Mandate versus championship: Vertical government intervention and diffusion of innovation in public services in authoritarian China." *Public Management Review* 16 (2014): 117–139.

Zhu, Xufeng. "Inter-regional diffusion of policy innovation in China: A comparative case study." *Asian Journal of Political Science* 25, no. 3 (2017): 266–286.

Zhu, Xufeng. "Executive entrepreneurship, career mobility and the transfer of policy paradigms." *Journal of Comparative Policy Analysis: Research and Practice* 20, no. 4 (2018): 354–369.

Zhu, Xufeng, and Guihua Bai. "Policy synthesis through regional experimentations: Comparative study of the new cooperative medical scheme in three Chinese provinces." *Journal of Comparative Policy Analysis: Research and Practice* 22, no. 4 (2020): 320–343.

Zhu, Xufeng, and Youlang Zhang. "Political mobility and dynamic diffusion of innovation: The spread of municipal pro-business administrative reform in China." *Journal of Public Administration Research and Theory* 26, no. 3 (2016): 535–551.

Zhu, Xufeng, and Youlang Zhang. "Diffusion of marketization innovation with administrative centralization in a multilevel system: Evidence from China." *Journal of Public Administration Research and Theory* 29, no. 1 (2019): 133–150.

Zhu, Xufeng, and Hui Zhao. "Recognition of innovation and diffusion of welfare policy: Alleviating urban poverty in Chinese cities during fiscal recentralization." *Governance* 31, no. 4 (2018): 721–739.

Zhu, Xufeng, and Hui Zhao. "Experimentalist governance with interactive central–local relations: Making new pension policies in China." *Policy Studies Journal* (Early View).

Zhu, Yapeng. "Policy entrepreneur, civic engagement and local policy innovation in China: Housing monetarisation reform in Guizhou province." *Australian Journal of Public Administration* 71, no. 2 (2012): 191–200.

Zhu, Yapeng. "Policy entrepreneurship, institutional constraints, and local policy innovation in China." *China Review* 13, no. 2 (2013): 97–122.

3 An institutional theory of local government innovativeness

The structural characteristics of the multi-level bureaucratic organisations in China

Like any organisations with a multi-regional governance form (i.e., the so-called "M-form") (Maskin, Quian, and Xu 2000; Qian, Roland, and Xu 2006), China's large-scale multi-level government system could be characterised by two key institutional parameters: the administrative ladder and the span of control. As Figure 3.1 shows, China has a five-tiered hierarchical government system, comprising national-, provincial-, city-, county-, and township-level governments (Zhang and Zhu 2020b). In China's unitary system, all governments except the national government are called "local governments". However, different levels of local governments tend to have varying resources and motivations. For instance, intuitively, higher-level governments tend to have more powers and resources than lower-level governments.

Moreover, the span of control in China's government system refers to the number of subordinate governments overseen by a superior government, such as the number of prefecture-level jurisdictions within a province or the number of county-level jurisdictions within a city. There is a great variation in the span of control between and within government tiers across China. For instance, according to the information provided by the Resource and Environment Data Cloud Platform (REDCP) of the Chinese Academy of Science (CAS), in 2015, the number of prefecture-level governments in each province ranged from 3 to 28 and the mean value was 14 in mainland China; the number of county-level governments in each prefecture-level city ranged from 0 to 25 and the mean value was 7. Notably, the current span of control along the administrative hierarchy in each region was mostly created by imperial dynasties (e.g., Yuan, Ming and Qing) (Sng et al. 2018). The span of control is stable in China because reform of the jurisdictional boundaries tends to be a politically complex and costly process,

An institutional theory of innovativeness 23

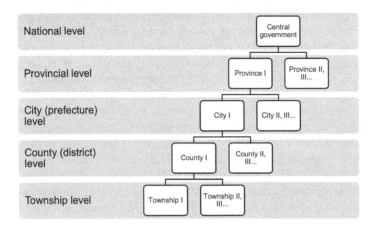

Figure 3.1 The hierarchy of the Chinese government system

which requires the compromise of multiple local stakeholders and the support of multiple tiers of higher-level governments (Zhang and Wu 2006).

This study's theoretical framework builds on insights from the analysis of the fundamental institutions of China's reforms and development in previous literature, such as decentralisation, promotion tournament, and career mobility (Xu 2011). The multi-level government system in China has three important institutional characteristics. First, local governments have certain policy autonomy. The decentralisation literature has shown that the vertical power structures (e.g., fiscal, administrative, or political powers) of countries are not uniformly centralised or decentralised (Cerniglia 2003; Schneider 2003). Typically, despite a unitary or federalist system, the vertical power structure in a country tends to be centralised in some policy areas and decentralised in other policy areas. In recent decades, many developed and developing countries have chosen to join the trend of decentralisation and various economic and administrative powers have been delegated to the local governments (Cai and Treisman 2005; Fan, Lin, and Treisman 2009). For instance, researchers have found that China's vertical power structure is a combination of fiscal and administrative semi-decentralisation and political centralisation (Montinola, Qian, and Weingast 1995; Xu 2011; Zhu and Zhang 2016). In other words, although the superior governments have the power of appointing subordinate leaders, the subordinate governments still have certain policy autonomy and resources to initiate innovative projects or adopt innovative policy instruments within the policy goals defined by the superior governments (Liu and Li 2016; Zhu and Zhang 2019).

Second, almost all government officials can only be promoted by level along the government hierarchy. The main leaders of each government tier are classified into five categories according to the government ranks: state leaders (e.g., president or prime minister), provincial leaders (e.g., provincial party chief or governor), city leaders (e.g., city party chief or mayor), county leaders (e.g., county party chief or county director), and township leaders (e.g., township party chief or township director) (Kou and Tsai 2014). Previous research shows that bureaucratic selection at multiple government tiers in China is shaped by both political connections (*guanxi*) and performance (Li and Zhou 2005; Serrato, Wang, and Zhang 2019; Yao and Zhang 2015). Particularly, despite abundant discussions of factional politics in China (Shih 2008; Shih, Adolph, and Liu 2012), connections and performance are complements rather than substitutes in China's political selection system (Jia, Kudamatsu, Seim 2015; Jiang 2018). Generally, Chinese government officials have strong incentives to adopt various policy tools to improve their performance and outperform their peers to get promoted (Xu 2011; Zhang and Zhu 2020b; Zhu 2014; Zhu and Zhang 2016).

Third, the superior governments often transfer subordinate leaders from one locality to another. In China, subnational leaders' careers tend to span multiple localities (Jiang and Mei 2020; Zhang and Gao 2008). The contemporary Chinese party-state inherited the geographic mobility system from ancient China, in which the rulers of empires designed the geographic mobility system to restrain local corruption and reduce local leaders' threats to the central government (Zhu 2018). However, in contemporary China, in addition to reducing local corruption and threats, the superior governments also transfer or rotate subordinate officials more frequently than previous empires did to increase local leaders' experience and capability, facilitate policy implementation, or promote innovative practices or cross-regional cooperation (Lieberthal and Oksenberg 1988; Thornton 2006; Xu 2011). The leadership transfer networks at each government tier are highly developed and tend to be significantly associated with innovation adoption in China (Yi, Berry, and Chen 2018; Zhu and Zhang 2016)

Local government innovativeness along the administrative hierarchy

Given China's institutional context, how does the administrative ladder affect local government innovativeness? Based on previous research, I point out that the administrative ladder could significantly shape local governments' resources and motivations regarding adopting or implementing local innovations. First, intuitively, higher-level governments could mobilise more resources than lower-level governments to introduce and support new reforms. In the past decades, China's decentralisation reforms mainly occurred between the national and provincial governments (Lin and Liu 2000). The Chinese style of "federalism" provides provincial governments with substantial policy autonomy and fiscal discretion (Montinola, Qian and Weingast 1995). Therefore, provincial governments could design various specific policies to suit local circumstances to support economic growth and demonstrate their presence to compete with the national government to attract the attention of lower-level governments (Zhu and Zhang 2019). By contrast, the lower-level governments' choices (e.g., city-, county-, or township-level governments) of policy instruments or reforms are often constrained by multiple superior-level governments. Moreover, during their innovative reforms, lower-level governments tend to suffer more than higher-level governments from limited time and resources and the potentially conflicting demands from multiple superior-level governments (Zhu and Zhang 2019). Therefore, as Figure 3.2 shows, I expect that governments' innovation resources are positively related to their positions in the administrative hierarchy.

The administrative ladder, however, might be negatively associated with local governments' innovation motivation. Recent research suggests that to maintain the survival of the regime and to mitigate the tradeoff between loyalty and efficiency in the promotion of local officials, Chinese central rulers strategically prioritise the observable performance indicators at the lower level of the government hierarchy and political connections or loyalty at the higher level of the government hierarchy (Landry, Lü and Duan 2018). The main reason is that the higher-level local officials are more important

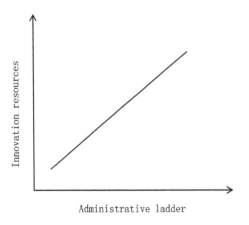

Figure 3.2 The relationship between administrative hierarchy and innovation resources

26 *An institutional theory of innovativeness*

for the survival of central rulers than the lower-level local officials because the former have higher political ranks and are more likely to be within the electorate of the national-level rulers. Based on a multi-level dataset of provincial, prefectural, and county-level party chiefs and government chiefs, Landy, Lü, and Duan (2018) found evidence consistent with the above argument. The above logic suggests that higher-level governments have fewer incentives to adopt innovative practices given the diminishing returns of the performance signals. By contrast, lower-level governments have stronger motivation to adopt innovative practices to send observable signals to their superiors (Zhu 2014). Therefore, as Figure 3.3 shows, I expect that governments' innovation motivations are negatively related to their positions in the administrative hierarchy.

Since both motivation and resources are necessary but insufficient conditions for generating innovations, as Figure 3.4 shows, combining the above two logics, this book proposes an inverse U-shaped relationship between the administrative ladder and local government innovativeness in China. Although provincial governments have substantial resources and policy autonomy regarding adopting innovative practices, they have far fewer incentives to adopt them because these innovations have negligible effects on their career promotion. Township governments, by contrast, have the strongest incentives to adopt innovations to create observable performance signals, but they have little policy autonomy or resources. Therefore, we expect that the county or city governments are more innovative compared to

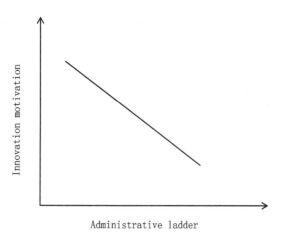

Figure 3.3 The relationship between administrative hierarchy and innovation motivation

An institutional theory of innovativeness 27

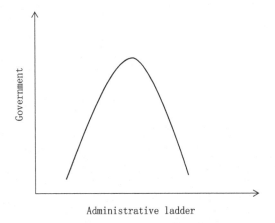

Figure 3.4 The relationship between administrative hierarchy and government innovativeness

the high-level or low-level governments given their less constrained innovation resources or innovation motivation.

Local government innovativeness along the span of control

As Figure 3.5 and Figure 3.6 show, there is a great variation in the span of control in China's multi-level government system. Theoretically, the span of control could affect local government innovativeness in three ways. First, as the span of control increases, the policy autonomy of local governments tends to increase, which could further increase the capability of local governments to innovate. Local policy autonomy is the premise of local innovative practices (Zhang and Zhu 2019). Since the early 20th century, numerous management scholars have argued that superiors should only oversee a certain number of subordinates in an organisation, given their resource and time limits, as the number of subordinates is positively associated with the difficulty of monitoring their behaviour (Gulick 1937; Ouchi and Dowling 1974; Woodward 1980). As the principal–agent research suggests, wide spans of control decrease the possibility of supervision and subordinates are more likely to have the discretion to develop responsibility for their actions (Brehm and Gates 1997). The implication of this logic is also reflected in McGregor's (1960) Theory X and Theory Y. In Theory X, superiors should minimise the span of control because they cannot trust their subordinates. In Theory Y, superiors should maximise the span of control to encourage subordinates to work independently and take on new responsibilities. In the

28 *An institutional theory of innovativeness*

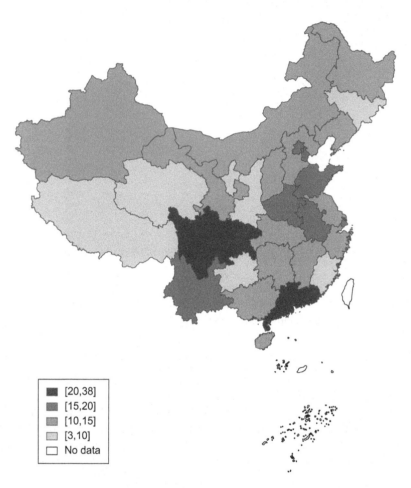

Figure 3.5 The number of prefecture-level regions governed by each province in China. (Note: This map is plotted based on the 2015 shapefile provided by the Resource and Environment Data Cloud Platform at the Chinese Academy of Sciences. No information is provided for Hong Kong, Macau, and Taiwan because they are not directly governed by the central government.)

context of China, the higher-level governments often face the monitoring and compliance challenges that come from decentralisation because it is costly for them to collect information about the lower-level governments (Anderson et al. 2019). Therefore, when the number of lower-level governments managed by a higher-level government increases, the higher-level

An institutional theory of innovativeness 29

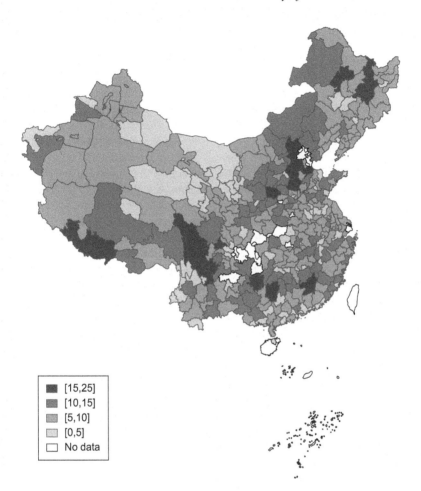

Figure 3.6 The number of county-level regions governed by each prefecture-level city in China. (Note: This map is plotted based on the 2015 shapefile provided by the Resource and Environment Data Cloud Platform at the Chinese Academy of Sciences. No information is provided for Hong Kong, Macau, and Taiwan because they are not directly governed by the central government. No information is provided for Beijing, Tianjin, Chongqing, and Shanghai because the county-level regions of these four major cities are actually towns. Dongguan, Zhongshan, Jiayuguan, Sansha, and Danzhou only govern towns. No information is provided for Tongren, Bijie, Karamay, and Sanya because these cities' geographic information is missing in the shapefile.)

governments' costs of monitoring lower-level governments also increase. Under these circumstances, lower-level governments' policy autonomy and the associated capability of adopting and implementing innovations will systematically increase as the risks of being intervened with or punished for innovative practices will decrease.

Second, as the span of control increases, the competition intensity between local leaders increases, which further increases their motivation to innovate and outperform their peers. In China's hierarchical career promotion system, the span of control in a jurisdiction is positively associated with the number of contestants (n), which is greater than the number of promotion events (k) if all these candidates are competent (Lü and Landry 2014). Therefore, the expected probability of promotion is the ratio of higher-level positions to lower-level leaders or k/n, which is negatively associated with the span of control or the intensity of career competition due to the smaller chance of being promoted to the higher positions. Admittedly, one may point out that subordinate leaders managed by the same superior government are not homogeneous (Zhang and Zhu 2020b). Some governments might simply be more politically important than other governments in terms of the political or economic status of their jurisdictions. For instance, in the case of Guangdong Province, provincial leaders are generally promoted from Shenzhen City or Guangzhou City (see, for instance, China Vitae: www.chinavitae.com/). However, notably, the leaders of Shenzhen City or Guangzhou City are generally promoted from other less important cities' leadership positions in Guangdong Province. Thus, the heterogeneity of the political or economic importance of subordinate governments only creates more ladders within a jurisdiction and this heterogeneity actually largely exists at each level of government and in each jurisdiction in China. Therefore, it does not change the outcome that more officials in the same government or party rank compete with each other when the span of control increases (Zhang and Zhu 2020b).

As competence or loyalty are difficult to observe, superiors often rely on easily visible indicators to select these candidates, such as local Gross Domestic Product (GDP), fiscal revenue, or innovative projects. Therefore, if one county from a city has adopted an innovation that could potentially improve its performance, then a strategically rational choice for other counties in that city would be to adopt their own innovations to avoid being outperformed and losing career opportunities (Zhang and Zhu 2020b; Zhu 2014). If all county leaders within a city reason similarly, then all counties' innovation motivations would be positively associated with the number of counties in the city. In other words, *ceteris paribus*, compared to a city with a smaller number of counties, a city with a greater number of counties would have more innovative county governments.

An institutional theory of innovativeness 31

Third, as the span of control increases, the opportunities for horizontal transfer tend to increase, which could further increase the opportunities for local governments to innovate. In China's government personnel system, local leaders cannot only be promoted or demoted vertically, but also be transferred horizontally. Intuitively, when, for instance, a city oversees a great number of counties, the promotion space for each county leader is small, but the opportunities of horizontal transfer are plenty. In fact, horizontal leadership transfer occurs frequently in China (Yi, Berry, and Chen 2018). Previous research based on cases in the United States and China has repeatedly found that compared to leaders who assumed their current positions from within, leaders from the outside are more likely to initiate innovations due to their accumulated professional knowledge in various socio-political environments and their ambition to deliver innovations to gain professional prestige in new positions (Bhatti, Olsen, and Pedersen 2011; Teodoro 2009, 2010, 2011; Zhu 2013; Zhu and Zhang 2016). Another possible reason is that local leaders who assumed their current positions from the outside tend to have more social networks across regions and are less likely to be constrained by stakeholders or organisational cultures in a jurisdiction and, thus, have more opportunities to adopt innovative practices (Bian 2002; Tsai and Dean 2014; Persson and Zhuravskaya 2016). Therefore, the span of control could systematically increase local government innovativeness through increasing the opportunities for horizontal leadership transfer.

As Figure 3.7 shows, the above analysis suggests that the span of control could increase local government innovativeness through three mechanisms, comprising increasing local innovation capability through increased policy autonomy, increasing local innovation motivation by producing greater competition intensity, and increasing local innovation opportunities by bringing more frequent transfer opportunities.

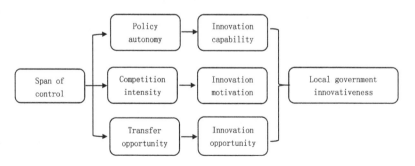

Figure 3.7 The theoretical framework

Bibliography

Anderson, Sarah E., Mark T. Buntaine, Mengdi Liu, and Bing Zhang. "Non-governmental monitoring of local governments increases compliance with central mandates: A national-scale field experiment in China." *American Journal of Political Science* 63, no. 3 (2019): 626–643.

Bhatti, Yosef, Asmus L. Olsen, and Lene Holm Pedersen. "Administrative professionals and the diffusion of innovations: The case of citizen service centres." *Public Administration* 89, no. 2 (2011): 577–594.

Bian, Yanjie. "Chinese social stratification and social mobility." *Annual Review of Sociology* 28, no. 1 (2002): 91–116.

Brehm, John O., and Scott Gates. *Working, shirking, and sabotage: Bureaucratic response to a democratic public*. Ann Arbor: University of Michigan Press, 1999.

Cai, Hongbin, and Daniel Treisman. "Does competition for capital discipline governments? Decentralization, globalization, and public policy." *American Economic Review* 95, no. 3 (2005): 817–830.

Cerniglia, Floriana. "Decentralization in the public sector: Quantitative aspects in federal and unitary countries." *Journal of Policy Modeling* 25, no. 8 (2003): 749–776.

Fan, C. Simon, Chen Lin, and Daniel Treisman. "Political decentralization and corruption: Evidence from around the world." *Journal of Public Economics* 93, no. 1–2 (2009): 14–34.

Gulick, Luther. "Notes on the theory of organization." *Classics of Organization Theory* 3, no. 1937 (1937): 87–95.

Jia, Ruixue, Masayuki Kudamatsu, and David Seim. "Political selection in China: The complementary roles of connections and performance." *Journal of the European Economic Association* 13, no. 4 (2015): 631–668.

Jiang, Junyan. "Making bureaucracy work: Patronage networks, performance incentives, and economic development in China." *American Journal of Political Science* 62, no. 4 (2018): 982–999.

Jiang, Junyan, and Yuan Mei. "Mandarins make markets: Leadership rotations and inter-provincial trade in China." *Journal of Development Economics* 147 (2020): 102524.

Kou, Chien-wen, and Wen-Hsuan Tsai. ""Sprinting with small steps" towards promotion: Solutions for the age dilemma in the CCP cadre appointment system." *The China Journal* 71 (2014): 153–171.

Landry, Pierre F., Xiaobo Lü, and Haiyan Duan. "Does performance matter? Evaluating political selection along the Chinese administrative ladder." *Comparative Political Studies* 51, no. 8 (2018): 1074–1105.

Li, Hongbin, and Li-An Zhou. "Political turnover and economic performance: The incentive role of personnel control in China." *Journal of Public Economics* 89, no. 9–10 (2005): 1743–1762.

Lieberthal, Kenneth, and Michel Oksenberg. *Policy making in China: Leaders, structures, and processes*. Princeton: Princeton University Press, 1988.

Lin, Justin Yifu, and Zhiqiang Liu. "Fiscal decentralization and economic growth in China." *Economic Development and Cultural Change* 49, no. 1 (2000): 1–21.

Liu, Wei, and Wenzhao Li. "Divergence and convergence in the diffusion of performance management in China." *Public Performance & Management Review* 39, no. 3 (2016): 630–654.

Lü, Xiaobo, and Pierre F. Landry. "Show me the money: Interjurisdiction political competition and fiscal extraction in China." *American Political Science Review* 108, no. 3 (2014): 706–722.

Maskin, Eric, Yingyi Qian, and Chenggang Xu. "Incentives, information, and organizational form." *The Review of Economic Studies* 67, no. 2 (2000): 359–378.

McGregor, Douglas. *The human side of enterprise* (Vol. 21, pp. 166–171). New York: McGraw-Hill, 1960.

Montinola, Gabriella, Yingyi Qian, and Barry R. Weingast. "Federalism, Chinese style: The political basis for economic success in China." *World Politics* 48, no. 1 (1995): 50–81.

Ouchi, William G., and John B. Dowling. "Defining the span of control." *Administrative Science Quarterly* 19, no. 3 (1974): 357–365.

Persson, P., and Zhuravskaya, E. "The limits of career concerns in federalism: Evidence from China." *Journal of the European Economic Association* 14, no. 2 (2016): 338–374.

Qian, Yingyi, Gerard Roland, and Chenggang Xu. "Coordination and experimentation in M-form and U-form organizations." *Journal of Political Economy* 114, no. 2 (2006): 366–402.

Qian, Yingyi, and Barry R. Weingast. "Federalism as a commitment to reserving market incentives." *Journal of Economic Perspectives* 11, no. 4 (1997): 83–92.

Schneider, Aaron. "Decentralization: Conceptualization and measurement." *Studies in Comparative International Development* 38, no. 3 (2003): 32–56.

Serrato, Juan Carlos Suárez Xiao Yu Wang, and Shuang Zhang. "The limits of meritocracy: Screening bureaucrats under imperfect verifiability." *Journal of Development Economics* 140 (2019): 223–241.

Shih, Victor, Christopher Adolph, and Mingxing Liu. "Getting ahead in the communist party: Explaining the advancement of central committee members in China." *American Political Science Review* 106, no. 1 (2012): 166–187.

Sng, Tuan Hwee, Pei Zhi Chia, Chen-Chieh Feng, and Yi-Chen Wang. "Are China's provincial boundaries misaligned?" *Applied Geography* 98 (2018): 52–65.

Teodoro, Manuel P. "Bureaucratic job mobility and the diffusion of innovations." *American Journal of Political Science* 53, no. 1 (2009): 175–189.

Teodoro, Manuel P. "Contingent professionalism: Bureaucratic mobility and the adoption of water conservation rates." *Journal of Public Administration Research and Theory* 20, no. 2 (2010): 437–459.

Teodoro, Manuel P. *Bureaucratic ambition: Careers, motives, and the innovative administrator*. Baltimore: The John Hopkins University Press, 2011.

Thornton, John L. "China's leadership gap." *Foreign Affairs* 85, no. 6 (2006): 133–140.

Tsai, Wen-Hsuan, and Nicola Dean. "Experimentation under hierarchy in local conditions: Cases of political reform in Guangdong and Sichuan, China." *China Quarterly* (2014): 339.

Woodward, Joan. *Industrial organization: Theory and practice*. 2nd ed. New York: Oxford University Press, 1980.

Xu, Chenggang. "The fundamental institutions of China's reforms and development." *Journal of Economic Literature* 49, no. 4 (2011): 1076–1151.

Yao, Yang, and Muyang Zhang. "Subnational leaders and economic growth: Evidence from Chinese cities." *Journal of Economic Growth* 20, no. 4 (2015): 405–436.

Yi, Hongtao, Frances Stokes Berry, and Wenna Chen. "Management innovation and policy diffusion through leadership transfer networks: An agent network diffusion model." *Journal of Public Administration Research and Theory* 28, no. 4 (2018): 457–474.

Zhang, Jun, and Yuan Gao. "Term limits and rotation of Chinese governors: Do they matter to economic growth?" *Journal of the Asia Pacific Economy* 13 (2008): 274–97.

Zhang, Jingxiang, and Fulong Wu. "China's changing economic governance: Administrative annexation and the reorganization of local governments in the Yangtze River Delta." *Regional Studies* 40, no. 1 (2006): 3–21.

Zhang, Youlang, and Xufeng Zhu. "Multiple mechanisms of policy diffusion in China." *Public Management Review* 21, no. 4 (2019): 495–514.

Zhang, Youlang, and Xufeng Zhu. "Career cohorts and inter-jurisdictional innovation diffusion: An empirical exploration in China." *International Public Management Journal* 23, no. 3 (2020b): 421–441.

Zhu, Xufeng. "Mandate versus championship: Vertical government intervention and diffusion of innovation in public services in authoritarian China." *Public Management Review* 16 (2014): 117–39.

Zhu, Xufeng. "Executive entrepreneurship, career mobility and the transfer of policy paradigms." *Journal of Comparative Policy Analysis: Research and Practice* 20, no. 4 (2018): 354–369.

Zhu, Xufeng, and Youlang Zhang. "Political mobility and dynamic diffusion of innovation: The spread of municipal pro-business administrative reform in China." *Journal of Public Administration Research and Theory* 26, no. 3 (2016): 535–551.

Zhu, Xufeng, and Youlang Zhang. "Diffusion of marketization innovation with administrative centralization in a multilevel system: Evidence from China." *Journal of Public Administration Research and Theory* 29, no. 1 (2019): 133–150.

Zhu, Yapeng. "Policy entrepreneurship, institutional constraints, and local policy innovation in China." *China Review* 13, no. 2 (2013): 97–122.

4 The Innovations and Excellence in Chinese Local Governance Awards Programme

This chapter provides a detailed review of the local governments' innovative practices in China by examining the IECLG Awards Programme from 2001 to 2015. Inspired by the IAGA Programme created by Harvard Kennedy School, the IECLG Awards Programme was an independent non-profit academic programme created in 2000 and first implemented in 2001 by the Comparative Politics and Economics Study Centre (CPESC) at the CCTB, the Comparative Study Centre of World Parties (CSCWP) at the CPSCCP, and the Centre for Chinese Government Innovations (CCGI) at Peking University (PKU). The chief leaders of this programme were Keping Yu, who was the director of the CPESC at the CCTB and the CCGI at PKU, and Changjiang Wang, who was the director of CSCWP at the CPSCCP.

The IECLG Awards Programme aimed to achieve six main goals (Yu 2014). First, it could help to identify the local innovations in institutional reforms, organisational changes, public services, and social management and promote "best practices" among local governments. Second, it could encourage and incentivise local governments to reform and improve existing governance systems. Third, it could provide a platform for scholars to develop government innovation theories based on scientific research. Fourth, researchers could develop a new performance evaluation system for local governments in China through the IECLG Awards Programme and provide intellectual supports for policymakers. Fifth, the IECLG team could exchange excellent practices and experiences from China with other countries in the world. Sixth, the IECLG Awards Programme could provide important experience and lessons for organising similar awards programmes in China.

The IECLG Awards Programme had several main characteristics (Yu 2009, 2010, 2014). First, the programme had its own independent evaluation procedures and standards and was not influenced by outside individuals or organisations; second, the programme was not profit-oriented and the IECLG team was responsible for all of the fees (including the monetary awards) related to the operation of the IECLG Awards Programme; third, the programme was implemented based on a set of procedures and standards that were created based on international experiences and the contextual characteristics in China; fourth, the programme was comprehensive as it involved a series of activities, such as research, award, and promotion, as well as training; finally, the programme was highly transparent in terms of application, field investigation, evaluation, and promotion.

Local governments were encouraged to submit applications to the IECLG team every two years (Yu 2006, 2007). However, every competitor had to make sure that its project met a series of conditions to qualify for review by the IECLG experts: first, the project had to be adopted by a local public service provider, comprising government branches and relevant social organisations; second, the project had to be adopted voluntarily rather than coercively; third, the goal of each project had to be the improvement of public benefits rather than private profits; fourth, the project had to be truly innovative rather than a simple replication of the activities of social groups or private organisations; fifth, the project had to produce observable public benefits and had to have been accepted by the public; sixth, the project must have been implemented for at least one year.

The IECLG team consisted of dozens of experts with backgrounds in political science, PA, PP, economics, and sociology (Yu 2014). These experts evaluated the local innovative practices according to six main indicators, comprising innovativeness, participation, effectiveness, importance, efficacy, and social acceptability. After the experts selected the candidate programmes, the list was posted on the website of the IECLG Awards Programme (www.chinainnovations.org/) and promoted by multiple influential websites, such as Xinhua News (www.xinhuanet.com/). The IECLG Awards Programme then sent multiple independent research teams to conduct field investigations and finally determined the finalists and winners of the awards. Every award winner (i.e., the winning organisation) received a cash prize of 50, 000 yuan.

I collected information on the IECLG Awards Programme from multiple sources for several years. I first obtained the full list of the first seven waves (2001–2014) of IECLG finalists and winners by directly contacting the staff of CCTB in 2014. I further collected the list of the eighth wave of IECLG finalists and winners from the public announcement of IECLG's official Wechat account and the PKU Research Centre for Chinese Politics (RCCP) in 2019. I also conducted face-to-face interviews with multiple experts involved in the IECLG Awards Programme to acquire first-hand knowledge about its operational and historical details. On this basis, I manually collected the detailed introduction of these IECLG finalists and winners from the *IECLG Case Study Reports* and government websites, and coded them following the practices of previous research on the same topic (Wu, Ma, and Yang 2013).

Table 4.1 shows the number of applicants, finalists, and winners of the IECLG Awards Programme. Since 2001, the IECLG Awards Programme was successfully conducted eight times. There were ten winners and around 12 finalists in each wave of the IECLG Awards Programme. There were around 300 applicants in each wave of the IECLG Awards Programme between 2001 and 2010. However, according to my interviews with the participants in the IECLG Awards Programme, after 2012, as the political environment became increasingly restrictive due to a series of centralised political campaigns initiated by President Xi Jinping, local officials had far less policy autonomy and took significantly higher political risks when adopting government innovations. Local officials were less willing to compete for the IECLG awards and there was a much smaller number of applicants in the eighth wave of the IECLG Awards Programme. Therefore, Yu Keping and other leaders of the IECLG Awards Programme terminated it in 2016. Nevertheless, among the 2007 applicants from 2001 to 2015, there were 80 winners and 99 finalists. Unfortunately, I could not retrieve the full list of those applicants as the IECLG team did not fully disclose the information when I requested it. Therefore, this study's analysis has focused on the winners and finalists rather than on the applicants.

Figure 4.1 shows the spatial distribution of the total number of IECLG winners and finalists (2001–2016) in each province, without distinguishing between the specific tiers of local innovation initiators within each province. Table 4.2 reports the number of IECLG winners and finalists by province and year. Most of the IECLG winners and finalists were located in the eastern and central areas of China. In general, the coastal local governments were more innovative than the inland local governments. The Zhejiang

Table 4.1 Description of the applicants, finalists, and winners of the IECLG Awards Programme

Year	Total applicants	Finalists	Winners
No.1 (2001–2002)	320	10	10
No.2 (2003–2004)	245	8	10
No.3 (2005–2006)	283	15	10
No.4 (2007–2008)	337	10	10
No.5 (2009–2010)	358	20	10
No.6 (2011–2012)	213	15	10
No.7 (2013–2014)	132	10	10
No.8 (2015–2016)	119	11	10
Total	2007	99	80

Note: As the IECLG team revised the number of the qualified applicants according to the criteria mentioned above, there were 320 applicants in the first IECLG Awards Programme rather than 325 applicants mentioned in Wu, Ma, and Yang (2013).

38 *The Innovations and Excellence Awards*

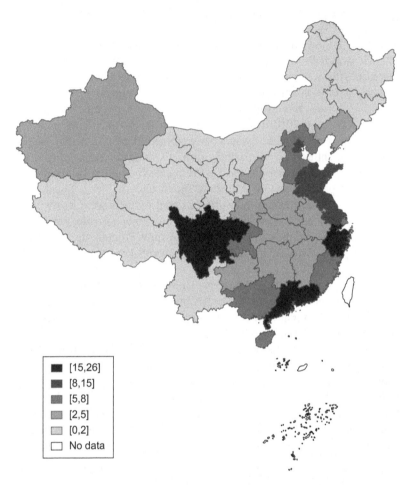

Figure 4.1 The total number of IECLG winners and finalists (2001–2016) in each province in China (Note: The map is plotted based on the 2015 shapefile provided by the Resource and Environment Data Cloud Platform at the Chinese Academy of Sciences. No information is provided for Hong Kong, Macau, and Taiwan because they are not directly governed by the central government.)

province could be thought of as China's California, as it provided the highest number (26) of IECLG award winners and finalists. Even during the recent COVID-19 pandemic, the Zhejiang province was the first adopter of multiple emergency management innovations, such as using the strengths of community-based organisations at multiple stages of COVID-19 response

Table 4.2 A list of the number of IECLG winners and finalists by province and year (2001–2015)

Province	2001	2003	2005	2007	2009	2011	2013	2015	Sub-total
Anhui	0	1	1	0	0	0	0	1	3
Beijing	0	2	2	1	2	2	1	1	9
Chongqing	0	0	2	0	1	1	1	1	6
Fujian	0	1	2	0	1	0	0	2	7
Gansu	0	0	0	0	0	0	0	0	0
Guangdong	2	1	1	2	3	4	2	3	18
Guangxi	1	2	1	1	0	1	0	0	6
Guizhou	1	0	0	0	1	0	1	0	3
Hainan	1	1	0	0	0	2	0	0	4
Hebei	1	2	2	0	1	1	0	0	7
Heilongjiang	0	1	0	1	0	0	0	0	2
Henan	1	0	1	0	0	0	1	1	4
Hubei	2	0	1	0	0	0	0	1	4
Hunan	1	0	1	0	0	0	1	0	3
Inner Mongolia	0	0	0	0	0	0	0	0	2
Jiangsu	2	0	2	1	4	1	2	1	12
Jiangxi	0	0	2	0	0	0	1	1	4
Jilin	0	1	0	0	0	0	1	0	2
Liaoning	0	0	1	0	1	1	0	1	4
Ningxia	0	0	0	0	0	0	0	1	2
Qinghai	0	0	0	0	0	0	0	0	0
Shaanxi	0	0	1	0	1	1	2	0	5
Shandong	0	1	0	2	2	1	2	1	9
Shanghai	2	0	1	2	1	1	0	1	8
Shanxi	0	0	0	0	0	0	0	1	1
Sichuan	2	2	2	3	2	2	2	2	17
Tianjin	0	0	1	0	0	1	0	0	2
Tibet	0	0	0	0	1	0	0	0	1
Xinjiang	1	0	0	1	1	0	0	0	3
Yunnan	0	0	0	0	1	0	0	0	2
Zhejiang	2	3	4	4	4	5	2	2	26
Sub-total	20	18	25	20	30	23	20	21	Total: 177

and when creating digital tracking platforms (Cheng et al. 2020). Moreover, Guangdong (18), and Sichuan (17) had the second and third highest number of innovative local governments, respectively. By contrast, Gansu (0) and Qinghai (0) had no local government winners of the IECLG awards.

Table 4.3 reports the specific initiators of IECLG winners and finalists. Various public organisations at each level of government have been actively involved in initiating government innovations, such as a people's congress, people's political consultative conference, department (bureau) of civil affairs, department (bureau) of justice, department (bureau) of public security, federation of trade unions, women's federation, legislative affairs office, financial affairs office, government service centre, administrative licensing centre, government audit office, and discipline inspection commission. Besides government organisations, the CCP committee and the Communist Youth League of China (CYLC) also played important roles in improving local public services. Notably, most of the IECLG winners and finalists were initiated by single public organisations. Only a few IECLG winners or finalist programmes were the result of horizontal collaborations between multiple government organisations, such as the "Minxin Network" initiated by the provincial-level discipline inspection commission, government supervision department, and government correction office in Liaoning Province. This finding suggests that there might be a "prisoner's dilemma" in collaborative governance between public organisations in China: public organisations are not willing to collaborate with each other to initiate collaborative innovations to avoid being viewed as a participant or imitator rather than a leader (Zhu 2014).

To further show the changes of the trends of local government innovations in China, I classified innovations in the Chinese public sector following Wu, Ma, and Yang (2013), comprising service innovation, technological innovation, management innovation, collaborative innovation, and governance innovation. Specifically, according to Wu, Ma, and Yang (2013), service innovation refers to "the supply of new services to new users, the delivery of existing services to new users or the supply of new services to existing users" (p. 350); technological innovation refers to "a change in service delivery technologies or arrangements" (p. 351); management innovation refers to "the restructuring of organisational structures and management processes and practices" (p. 351); collaborative innovation refers to "boundary-spanning activities in the process of service delivery and management (for example, alliances, partnerships, collaborations and networking)" (p. 352); governance innovation refers to "new approaches and practices that aim to manage democratic institutions, trigger citizen participation and fight corruption" (p. 352). The coding details of all IECLG winners and finalists are provided in the appendix. Notably, many IECLG

Table 4.3 The specific initiators of IECLG winners and finalists (2001–2015)

Innovation initiator	Frequency
Provincial-level	
Provincial-level government	3
Provincial committee of the CYLC's Chunhui action development centre	1
Provincial-level discipline inspection commission, government supervision department, and government correction office	1
Provincial-level department of civil affairs	2
Provincial-level department of environmental protection	1
Provincial-level department of justice	1
Provincial-level department of public security	1
Provincial-level disabled persons' federation	1
Provincial-level federation of trade unions	2
Provincial-level government service centre	2
Provincial-level financial affairs office	1
Provincial-level office of organisation and bianzhi committee	2
Provincial-level people's congress	1
Provincial-level public security frontier corps	1
Provincial-level women's federation	1
City-level	
City government	28
City government and CCP committee	2
City-level bureau of civil affairs	2
City-level bureau of finance	1
City-level bureau of human resources and social insurance	1
City-level bureau of letters and complaints	1
City-level bureau of social organisation management	1
City-level bureau of supervision	1
City-level federation of trade unions	3
City-level health and family planning commission	1
City-level legislative affairs office	3
City-level office of comprehensive evaluation committee	1
City-level office of legal affairs	2
City-level organisation department of CCP committee	2
City-level people's congress standing committee	1
City-level people's political consultative conference	1
City-level political and legal committee	3
City-level publicity department of the CCP committee	2
City-level social work committee	1
County government	1
District government	7
District government and CCP committee	1
District-level administrative licensing centre	2
District-level bureau of civil affairs	1
District-level federation of trade unions	1
District-level office of the committee of comprehensive management of social security	1
District-level people's procuratorate	1
County-level	
City government	9

(*Continued*)

Table 4.3 Continued

Innovation initiator	Frequency
City government and CCP committee	2
City-level CCP committee	2
City-level federation of trade unions	1
City-level people's congress standing committee	1
County government	21
County government and CCP committee	5
County-level CCP committee	1
County-level discipline inspection commission	2
County-level external aid project office	1
County-level organisation department of CCP committee	1
District government	18
District government and CCP committee	2
District-level CCP committee	2
District-level bureau of finance	1
District-level government audit office	1
Township government	2
Township-level office of street affairs	2
Township-level people's congress	1
Township-level Township government	4
Township government and CCP committee	1
Township-level office of street affairs	3
Township-level women's federation	1

Note: The township-level governments in municipalities directly under the central government are listed as county-level governments; the county-level or district-level governments in municipalities directly under the central government are listed as city-level governments; there are multiple county-level city governments in the sample. The reform of the collective forestry right institution initiated by National Forestry Administration (NFA) of China (2011) and the volunteering project of publicising climate disaster responses initiated by National Meteorological Administration (NMA) of China (2011) were listed as the finalists in the sixth IECLG competition and are not listed here as local innovations.

winners and finalists may involve multiple types of innovation simultaneously. Nevertheless, I classified each IECLG winner or finalist into only one innovation category because the purpose of my classification was only to capture the main characteristic of each IECLG winner or finalist in order to show the general trend of innovation typology in China.

As Table 4.2 shows, among the 179 local IECLG winner and finalists, there were 15 collaborative innovations (e.g., the privatisation of public welfare undertakings in Ganchahe Town, Shucheng County, Anhui Province), 50 governance innovations (e.g., the creation of People's Political Consultative Conference's Deliberation Room in Anyang City, Henan Province), 46 management innovations (e.g., the administrative reform and process reengineering in Zhangpu Town, Kunshan City, Jiangsu Province), 58 service innovations (e.g., the village-level public service and social management

Table 4.4 The typology and distribution of the IECLG winners and finalists (2001–2015)

	2001	2003	2005	2007	2009	2011	2013	2015	Total
Collaborative innovation	1	3	0	2	2	5	2	0	15
Governance innovation	11	5	8	2	8	6	1	9	50
Management innovation	4	2	7	6	6	9	7	5	46
Service innovation	3	8	8	8	12	6	8	5	58
Technological innovation	1	0	2	2	2	1	2	2	12
Total	20	18	25	20	30	27	20	21	179

reform in Chengdu City, Sichuan Province), and 12 technological innovations (e.g., the construction of the Citizen Health Information System in Xiamen City, Fujian Province).

Given China's rapid economic growth in recent years, it is intuitive to expect that local governments took advantage of management innovations and service innovations to improve business environments and attract investments. Moreover, Chinese society has suffered from various negative consequences caused by its overemphasis on economic development, such environmental pollution, economic inequality, and various social or cultural controversies, which probably motivated local governments to adopt the governance innovations to maintain social stability (Cai 2010; Steinhardt and Wu 2016; Wang 2014).

Bibliography

Cai, Yongshun. *Collective resistance in China: Why popular protests succeed or fail*. Stanford: Stanford University Press, 2010.

Cheng, Yuan, Jianxing Yu, Yongdong Shen, and Biao Huang. "Coproducing responses to COVID-19 with community-based organizations: Lessons from Zhejiang Province, China." *Public Administration Review* 80, no. 5 (2020): 866–873.

Steinhardt, H. Christoph, and Fengshi Wu. "In the name of the public: Environmental protest and the changing landscape of popular contention in China." *The China Journal* 75, no. 1 (2016): 61–82.

Wang, Yuhua. "Empowering the police: How the Chinese Communist Party manages its coercive leaders." *The China Quarterly* 219, no. 9 (2014): 625–648.

Wu, Jiannan, Liang Ma, and Yuqian Yang. "Innovation in the Chinese public sector: Typology and distribution." *Public Administration* 91, no. 2 (2013): 347–365.

Yu, Keping (eds). *Innovations and excellence in Chinese local governance: Case study reports (2003–2004)*. Beijing, China: Peking University Press (in Chinese), 2006.

Yu, Keping (eds). *Innovations and excellence in Chinese local governance: Case study reports (2005–2006)*. Beijing, China: Peking University Press (in Chinese), 2007.

Yu, Keping (eds). *Innovations and excellence in Chinese local governance: Case study reports (2007–2008)*. Beijing, China: Peking University Press (in Chinese), 2009.

Yu, Keping (eds). *Innovations and excellence in Chinese local governance: Case study reports (2009–2010)*. Beijing, China: Peking University Press (in Chinese), 2010.

Yu, Keping (eds). *Innovations and excellence in Chinese local governance: Case study reports (2011–2012)*. Beijing, China: Peking University Press (in Chinese), 2014.

Zhu, Xufeng. "Mandate versus championship: Vertical government intervention and diffusion of innovation in public services in authoritarian China." *Public Management Review* 16 (2014): 117–139.

5 An empirical investigation based on the IECLG Awards Programme

The measurement of local government innovativeness

Empirically, how do we operationalise government innovativeness? Government innovativeness is an organisational-level attribute rather than an innovation-level attribute. The literature review in Chapter 2 shows that most previous studies on local government innovation in China have mainly focused on one or several specific innovations and therefore largely only contribute to our understanding of specific innovation cases rather than the organisational-level innovativeness of local governments. Zhao (2012) provided the only study that has attempted to empirically analyse the determinants of government innovativeness in China. However, as mentioned before, Zhao's measure of government innovativeness includes innovations initiated by multiple tiers of the government, which may induce significant measurement errors.

The literature on government innovativeness has provided two main approaches to operationalising this concept. The first potential measurement is public managers' perceived organisational innovativeness. For instance, in a survey of English local governments, Walker, Berry, and Avellaneda (2015) asked the respondents to rate the degree of organisational innovativeness in terms of five types of innovations, comprising service innovation, marketisation innovation, organisation innovation, technological innovation, and ancillary innovation. All questions were in the form of Likert scales and the perceived measures were summed up to account for the "total innovation" of a public organisation. Nevertheless, these public manager respondents are not high-ranking government leaders, which means they are probably not the ones who really make the decisions regarding government innovation. The second potential measurement is a count of government innovations recognised by influential academic institutions. For instance, Ma (2017) used the raw number of state government innovations recognised by the Innovations in American Government Awards

(IAGA) Programme (i.e., the IAGA finalists and winners) to construct state government innovativeness.

In general, compared to the first measurement approach, the second measurement approach is preferred in terms of objectivity, replicability, and data availability. Intuitively, different from public managers' perceived organisational innovativeness, a count of government innovations recognised by influential academic institutions are less likely to be influenced by subjective biases and, therefore, more replicable. Moreover, large-scale surveys of high-ranking government leaders are often difficult for scholars. By contrast, information on the government innovation programmes led by nonprofit academic organisations is often easily available through websites, case reports, journal articles, or books.

In this study, I followed Ma (2017) and used the number of IECLG winners *and* finalists in a jurisdiction to measure the innovativeness of local governments in China. For instance, a provincial government's innovativeness is measured by the number of provincial-level IECLG winners and finalists in a province in a year; a city government's innovativeness is measured by the number of city-level IECLG winners and finalists in a city in a year. There are two reasons for this decision. First, the novelty and usefulness of the IECLG winners and finalists are widely recognised by practitioners and scholars (Almén 2016; Wang and Guo 2015; Wu, Ma, and Yang 2013), which could significantly improve our confidence in the validity of this measurement. Second, choosing this previously used measurement could facilitate cross-country comparisons between the findings based on China and other countries, such as the United States (Ma 2017) and Canada (Bernier, Hafsi and Deschamps 2015).

The administrative hierarchy and local government innovativeness

Despite the theoretical analysis in Chapter 3, it is difficult to determine the causal effects of administrative hierarchy on government innovativeness empirically. Ideally, to identify the theorised causal effects empirically, we may compare a high-level government's innovativeness with a low-level government's innovativeness, assuming other things being equal. However, other things are often not equal, and governments tend to be heterogeneous in terms of organisational and contextual characteristics. Governments, at different tiers and locations, often have varying human and fiscal resources and their jurisdictions have historical, cultural, political, economic, demographic, and geographical advantages or disadvantages that could have heterogeneous effects on their innovation motivations, innovation obstacles, and innovation resources (Berry and Berry 1990; Mohr 1969). Therefore, directly comparing a high-level government's innovativeness with a low-level government's innovativeness will disregard the potential confounding effects of historical, cultural, geographical, and other unobserved or unknown factors.

Alternatively, we may take advantage of potential natural shocks that only exogenously increase or decrease the administrative ranking of a local government in the observation period and hopefully, rule out the potential confounding effects of other factors. For instance, if a county-level government was suddenly promoted to a prefecture-level government, we may take advantage of this opportunity to compare government innovativeness before and after this change. Nevertheless, after a detailed examination of the 177 local IECLG innovation cases in the sample, I found no changes in the administrative ranking of these local governments across the observation period (2001–2016). In fact, the changes in local political, economic or environmental conditions in a jurisdiction are often the main determinants of change in the administrative hierarchy of a local government (Chung and Lam 2004). Under these circumstances, even if we observe a change of local government innovativeness, it is still difficult to distinguish between the effects of innovation resources and innovation motivations determined by administrative hierarchy (i.e., my theorised effect in Chapter 3) and the effects of the innovation needs caused by the previous environmental changes. Nevertheless, the empirical difficulty in identifying the causal effect of administrative hierarchy does not decrease the theoretical importance of this important structural characteristic of government organisations in explaining local government innovativeness.

Therefore, in this section, I present only a preliminary and exploratory empirical analysis to provide suggestive evidence to support the theorised effects of administrative hierarchy on local government innovativeness, thus providing a scholarly basis for future theoretical or empirical research on the same topic. According to the theoretical analysis in Chapter 3, I expected that compared to the middle-level governments (i.e., the city- or county-level governments), the provincial-level governments and township-level governments would have been less likely to apply for recognition and would have received less recognition (i.e., fewer IECLG winners or finalists) by the IECLG programme. Unfortunately, information on the applicants is not available and only the results of the IECLG winners or finalists can be reported here.

Table 5.1 shows the distribution of IECLG winners and finalists across government tiers and years. On average, there were around three winners and finalists at the provincial level, around nine winners and finalists at the city or county level, and around just one winner or finalist at the township level in each wave of the IECLG Awards Programme. Among the 177

48 An empirical investigation on the awards

Table 5.1 The number of IECLG winners and finalists across government tiers and years

	2001	2003	2005	2007	2009	2011	2013	2015	Total
Provincial level	0	1	3	3	2	5	5	2	21
City level	8	10	8	6	15	10	4	11	72
County level	10	5	13	11	12	7	9	8	75
Township level	2	2	1	0	1	1	2	0	9
Total	20	18	25	20	30	25	20	21	177

Note: The government tier is coded according to the administrative ranking of the innovation adopters. The reform of collective forestry right institution initiated by the National Forestry Administration (NFA) of China (2011) and the volunteering project of publicising climate disaster responses initiated by the National Meteorological Administration (NMA) of China (2011) were listed as the finalists in the sixth IECLG competition and are not listed here as provincial-level innovations.

winners and finalists in the eight waves of the IECLG Awards Programme, 21 belonged to the provincial-level governments, 73 belonged to the city-level governments, 75 belonged to the county-level governments, and only nine belonged to the township level governments. There are nearly 40,000 township governments in China, but they have the smallest number of IECLG winners or finalists. These findings seem to suggest that city- or county-level governments are the dominant innovators in China's public sector.

Admittedly, one may argue that in China's hierarchical government system, there are more lower-level governments than higher-level governments and therefore, Table 5.1 seems to suggest that provincial governments are actually much more innovative than city-level governments and we may infer that the administrative ladder is positively associated with government innovativeness. However, it should be noted that the range of provincial-level winners and finalists are between only zero and five and the range of city or county-level winners and finalists are between four and 15 or five and 13, which suggests that the number of city- or county-level winners and finalists are more likely to be restricted by the total number of winners and finalists in the IECLG programme in each year. In other words, the innovativeness of city- or county-level governments is probably much higher than the number of winners and finalists suggests. Moreover, provincial governments face fewer competitors than city governments and it is probably easier for a provincial government to successfully become a winner or a finalist if it applied for the IECLG Awards Programme.

Table 5.2 further presents the typology and distribution of the IECLG winners and finalists across government tiers. Provincial-, city-, and

Table 5.2 The typology and distribution of the IECLG winners and finalists across government tiers

	Collaborative innovation	Governance innovation	Management innovation	Service innovation	Technological innovation	Total
Provincial level	1	3	4	10	3	21
City level	4	14	27	21	7	73
County level	6	30	14	22	2	74
Township level	2	3	1	3	0	9
Total	14	50	46	57	12	177

county-level governments have initiated collaborative innovations, governance innovations, management innovations, and service innovations. Most types of innovation were adopted by the city and county governments. Probably due to the limits of resources and staff, China's township-level governments do not have IECLG winners or finalists that involve technological innovations.

Despite the lack of solid empirical evidence, I believe the theorisation and preliminary analysis of the relationship between administrative hierarchy and government innovativeness are meaningful in two ways. On the theoretical side, by developing testable theoretical predictions, the theoretical framework paves the way for research to further explore the potential effects of structural characteristics of China's government organisations on their innovativeness. On the practical side, analysing the correlations between administrative hierarchy and government innovativeness could provide useful guidance for government policymakers to improve resource allocation for different tiers of governments when they try to create innovative public organisations or encourage or promote policy innovations.

The span of control and local government innovativeness

We empirically operationalised the span of control in China's government system by measuring it with the number of subordinate governments within a jurisdiction. Therefore, for the provincial governments, there is no variation in span of control because they are all overseen by the same central government. By contrast, the span of control within a province could be measured with the number of prefecture-level cities in it. Similarly, the span of control within a city could be measured with the number of counties in it, and the span of control within a county could be measured with the number of townships in it. Nevertheless, due to the commonly recognised difficulty in collecting county-level and township-level governmental and contextual information in China, this book mainly focuses on how the span of control within a province affects the innovativeness of city governments to test the key theoretical expectation put forward in Chapter 3.

Notably, the span of control within a province or a city was generally stable in China during the observation period. Since 2004, although the reform of "Province-Managing-County" (PMC) has become increasingly popular in China (Liu and Alm 2016), it is mainly about enabling county-level governments to directly receive funding from the provincial-level governments and augmenting local finance capacity (Huang et al. 2017). The reformed counties are still mainly overseen by the city-level governments in terms of personnel mobility and other administrative affairs, and county-level leaders still need to compete with

An empirical investigation on the awards 51

each other to become city-level leaders in the government hierarchy. Therefore, this PMC reform does not really challenge the theoretical mechanisms of span of control discussed in Chapter 3.

Figure 5.1 shows that there is a positive correlation between the number of city-level IECLG winners and finalists and the number of city-level regions in each province. Figure 5.2 shows a positive correlation between the number of IECLG winners and finalists and the number of county-level

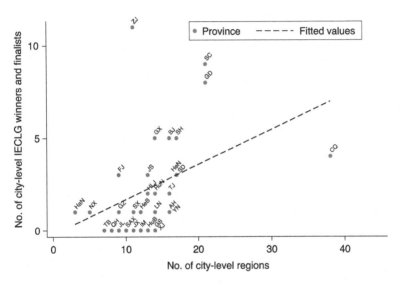

Figure 5.1 The correlational relationship between the number of city-level IECLG winners and finalists and the number of city-level regions in each province from 2001 to 2015 (Note: BJ="Beijing", TJ="Tianjin", HeB="Hebei", SX="Shanxi", IM="Inner Mongolia", LN="Liaoning", JL="Jilin", HLJ="Heilongjiang", SH="Shanghai", JS="Jiangsu", ZJ="Zhejiang", AH="Anhui", FJ="Fujian", JX="Jiangxi", SD="Shandong", HeN="Henan", HuB="Hubei", HuN="Hunan", GD="Guangdong", GX="Guangxi", HaN="Hainan", CQ="Chongqing", SC="Sichuan", GZ="Guizhou", YN="Yunnan", TB="Tibet", SAX="Shaanxi", GS="Gansu", QH="Qinghai", NX="Ningxia", XJ="Xinjiang". Shandong and Henan are overlapped because both of them have 17 city-level regions and 3 city-level IECLG winners and finalists. Similarly, Gansu and Xinjiang have 14 city-level regions and 0 city-level IECLG winners and finalists; Anhui and Yunnan have 16 city-level regions and 1 city-level IECLG winners and finalists. No information is provided for Hong Kong, Macau, and Taiwan because they are not directly governed by the central government and not covered by the IECLG programme.)

52 *An empirical investigation on the awards*

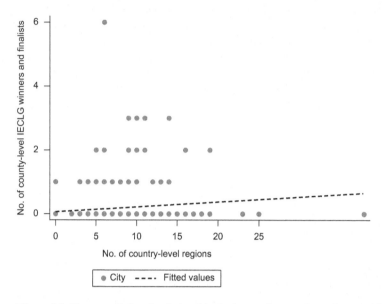

Figure 5.2 The correlational relationship between the number of county-level IECLG winners and finalists and the number of county-level regions in each prefecture-level city from 2001 to 2015 (Note: No information is provided for Hong Kong, Macau, and Taiwan because they are not directly governed by the central government and not covered by the IECLG programme.)

regions in each prefecture-level city across the observation period. These findings provide preliminary evidence suggesting that the span of control positively correlates with local government innovativeness. Admittedly, one may argue that, for instance, a province with more city-level governments may simply have more opportunities to win the city-level IECLG prizes. However, as reported before, the annual number of IECLG winners and finalists ranges from 18 to 30. Most provinces have only one or even no city-level IECLG winners or finalists in a year and the number of city-level governments should not be able to directly improve the odds of winning the IECLG prizes. In fact, among the 248 province-years in the research sample, only 13 province-years have two IECLG winners or finalists. Similarly, most of the cities only have zero or one county-level IECLG winner or finalist in a year. Among the 1,700 city-year observations, only Ankang City of Shaanxi Province has one county-level winner and one county-level finalist in 2013. Therefore, those IECLG winners or finalists are probably truly more innovative than other applicants. To further explore the potential theoretical relationships between the span of control and local government innovativeness, I provide more detailed provincial-level and city-level statistical analyses in this section.

Provincial-level statistical analysis

I evaluated the effect of span of control by first conducting the provincial-level panel data analysis. The provincial-level research sample includes 31 provincial-level regions directly managed by the central government. The observation period ranges from 2001 to 2015, in which the IECLG Awards Programme was held biennially eight times, including 2001, 2003, 2005, 2007, 2009, 2011, 2013, and 2015. The unit of analysis is province-year, and the statistical analyses include 248 observations.

The definition of all variables, along with detailed measurements and data sources are provided in Table 5.3. The dependent variable, the city-level government innovativeness of a province, is measured by the number of city-level IECLG winners or finalists in a province in a year. Among the 248 provincial-level observations, 188 of them have no IECLG winners or finalists; 47 observations have one IECLG winner or finalist; and only 13 province-years have two IECLG winners or finalists. Although it is reasonable to argue that governments with two IECLG winners or finalists are more innovative than governments with one IECLG winner or finalist, we cannot directly establish a precise numerical relationship between them and argue that the former's innovativeness is two times that of the latter's innovativeness. More importantly, it is unrealistic to assert that governments without an IECLG winner or finalist simply have no innovativeness. Therefore, we should treat the dependent variable as an ordinal variable rather than an interval-scale or ratio-scale variable. I used ordered logit models rather than Poisson or Negative Binomial models to conduct the statistical analyses.

One may argue that we should further divide the dependent variable by the number of city-level governments in a province to account for the possibility that provinces with more city-level governments simply have more opportunities to win in the IECLG Awards Programme. However, given the ordinal nature of our dependent variable and that most of the values are zeros, the approach of dividing the dependent variable by the number of city-level governments in a province should not change the results of the ordered logit models. Due to the competitive innovation learning patterns found in previous research (Zhu 2014), I argue that the number of city-level IECLG winners or finalists should be able to represent the aggregate level of city-level government innovativeness in a province. I collect the information on the dependent variable from the websites of the CCTB, CPS, and PKU.

Table 5.3 Variables, measures, and data sources for provincial-level analysis (2001–2015)

Variables	Description of Measurement	Sources
Dependent variable		
Number of county-level IECLG winners or finalists	Number of city-level IECLG winners or finalists in a province in a year	The websites of the Central Compilation and Translation Bureau, Party School of the CPC Central Committee, and Peking University
Independent variable		
Number of city-level regions	The number of city-level regions in a province in a year	The website of the National Bureau of Statistics of China (www.data.stats.gov.cn)
Control variables		
Fiscal dependence (%)	The extent to which a provincial government depends on the central government for fiscal resources. This variable is calculated by $$\frac{\text{Expenditure}_{i,t-1} - \text{Revenue}_{i,t-1}}{\text{Expenditure}_{i,t-1}} \times 100$$ (Zhang 2015)	National Bureau of Statistics of China (www.data.stats.gov.cn)
Service sector (%)	Percentage of the service sector in the province's GDP in the previous year	
Log (Foreign investment)	Natural logarithm of contracted investments (million dollars) in a province in the previous year	
Log (GDP)	Natural logarithm of GDP (100 million yuan) in a province in the previous year	
Log (Population)	Natural logarithm of the population size in a province in the previous year	
Log (Area size)	Natural logarithm of the area size (square kilometres) of a province	
Minority autonomous region	Dummy=1 if the observed unit is a provincial-level minority autonomous region, such as Guangxi, Tibet, Xinjiang, Ningxia or Inner Mongolia	
Municipality directly under the central government	Dummy=1 if the observed unit is a provincial-level municipality directly under the central government, such as Beijing, Tianjin, Shanghai, and Chongqing	

The key independent variable, the span of control in a province, is measured by the number of city-level administrative regions in a province in a year. I collected this data from the website of the National Bureau of Statistics (NBS) of China (www.data.stats.gov.cn). According to the theoretical analyses in Chapter 3, I expected that the independent variable positively predicts the dependent variable.

I also include a number of variables to control for the characteristics of each province in our analysis. First, a control for fiscal dependence, which refers to the extent to which a provincial government depends on the central government for fiscal resources. This variable is calculated by $\frac{\text{Expenditure}_{i,t-1} - \text{Revenue}_{i,t-1}}{\text{Expenditure}_{i,t-1}} \times 100$ for province i in year t (Zhang 2015). I expect fiscal dependence to negatively predict government innovativeness because governments with a higher level of fiscal dependence tend to have a lower level of policy autonomy and resources for adopting innovative policies or projects. Second, the percentage of the service sector in the province's GDP is included in the model because, given the prominent contribution of the service sector to local tax and employment status, a province with a more developed service sector may have more active citizens and enterprises that demand better and innovative public services (Berry and Berry 1990; Zhu and Zhang 2016).

Third, I include the size of the contracted investments (in millions of dollars) in a province in a year in the models to account for the potential effects of economic openness. I took the natural logarithm of this variable to capture the potential relative change (e.g., a per cent change) rather than absolute change (e.g., a unit change) caused by the monetary variables with a highly skewed distribution. Fourth, similarly, I included the natural logarithm of GDP in a province in a year in the models to account for the potential effect of economic size on government innovativeness. Fifth, I included the natural logarithm of the population size in a province in a year in the models as a city with a bigger population size may have more diverse demands for public service innovations. Sixth, I included the natural logarithm of the area size (square kilometres) of a province in the models because the area size of a province may determine both its number of city-level regions (i.e., the independent variable) and the diversity or intensity of public demands for government innovativeness (i.e., the dependent variable).

Seventh, I further controlled for the administrative categories of each provincial-level unit, comprising normal provinces, minority autonomous regions, and municipalities directly under the central government. Specifically, I created a dummy equal to one if the observed unit is a provincial-level minority autonomous region, such as Guangxi, Tibet, Xinjiang,

Ningxia, or Inner Mongolia; I also created a dummy equal to one if the observed unit is a provincial-level municipality directly under the central government, such as Beijing, Tianjin, Shanghai, and Chongqing. In other words, the reference category is the normal provinces.

Finally, in the statistical models, I lagged the time-varying provincial-level control variables by one year to avoid reverse causation (Zhu and Zhang 2016, 2019). My analysis also accounted for the potential temporal dependence problem by adding the year fixed effects.

Table 5.4 reports the summary statistics of all variables in the provincial-level statistical analysis (2001–2015). On average, a province has 0.29 city-level IECLG winners or finalists in a year; the minimum value is zero and the maximum value is two. Moreover, on average, the number of city-level regions in a province has a mean value of 13.69 and ranges from two (i.e., Hainan Province) to 40 (i.e., Chongqing Municipality). The fiscal dependence on the central government has an average value of 49.13 and ranges from 6.23 to 94.70. The per cent of the service sector in the GDP of a province has an average value of 41.42 and ranges from 28.30 to 77.95. The

Table 5.4 Summary statistics of the variables for provincial-level analysis (2001–2015)

Variable	Observations	Mean	Standard. Deviation	Minimum Value	Maximum Value
Number of city-level IECLG winners or finalists	248	0.29	0.56	0.00	2.00
Number of city-level regions	248	13.69	6.34	2.00	40.00
Fiscal dependence (%)	248	49.13	20.23	6.23	94.70
Service sector (%)	248	41.42	7.94	28.30	77.95
Log (Foreign investment)	248	10.01	1.62	5.80	13.48
Log (GDP)	248	8.69	1.21	4.77	11.12
Log (Population)	248	17.27	0.87	14.76	18.49
Log (Area size)	248	11.48	1.15	8.75	13.39
Minority autonomous region	248	0.16	0.37	0.00	1.00
Municipality directly under the central government	248	0.13	0.34	0.00	1.00

natural logarithm of foreign investments (million dollars) has an average value of 10.01 and ranges from 5.80 to 13.48. The natural logarithm of GDP (100 million yuan) has an average value of 8.69 and ranges from 4.77 to 11.12. The natural logarithm of population size has an average value of 17.27 and ranges from 14.76 to 18.49. The natural logarithm of area size (square kilometres) has an average value of 11.48 and ranges from 8.75 to 13.39. A total of 16 per cent of the observations in the sample belong to the minority autonomous regions, and 13 per cent of the observations in the sample belong to the municipalities directly under the central government; in other words, 71 per cent of the observations in the sample belong to the normal provinces.

Table 5.5 reports the estimates of city-level IECLG winners or finalists in a province. Specifically, Table 5.5 presents specifications with controls only alongside specifications that include the key independent variable, the number of city-level regions in a province. The key independent variable improves the Pseudo R-squared by 0.09, indicating that the number of city-level IECLG winners or finalists in a province vary significantly by the number of city-level regions in a province. The odds ratio for one unit increases in each independent variable and t-statistics in parentheses are provided. The two-tailed p values are shown in separate columns.

The results are consistent with previous hypotheses regarding the relationship between the span of control and local government innovativeness. Specifically, as Model 2 shows, one more city-level region in a province is associated with an 8.8 per cent increase in the odds of having one more IECLG winner or finalist in a province in a year, with a p-value of 0.041. According to the theoretical analysis in the previous chapter, this evidence suggests that the number of city-level regions in a province may be positively related to city governments' policy autonomy and city leaders' competition intensity and horizontal transfer opportunities within a province, thus increasing the overall city-level government innovativeness in a province. Nevertheless, we do not have strong evidence to support the potential effects of other provincial-level socioeconomic or geographic characteristics on city-level government innovativeness.

City-level statistical analysis

To further verify the effect of the span of control on local government innovativeness, I also conducted a city-level panel data analysis. The research sample included around 270 cities across the observation period. Since most of the cities in the sample have the missing value problem from 2001 to 2003 and the time-varying city-level variables need to be lagged by one year to avoid reverse causation, the observation period ranged from 2005 to 2015,

Table 5.5 Provincial-level analysis results (2001–2015)

	Model 1		Model 2	
	Odds Ratio (t-statistic)	p-value	Odds Ratio (t-statistic)	p-value
Number of city-level regions			1.088 (2.04)	0.041
Fiscal dependence (%)	0.950 (−0.91)	0.365	0.948 (−0.96)	0.339
Service sector (%)	0.997 (−0.12)	0.908	1.012 (0.42)	0.675
Log (Foreign investment)	1.291 (0.41)	0.684	1.229 (0.34)	0.734
Log (GDP)	0.340 (−1.01)	0.313	0.464 (−0.66)	0.509
Log (Population)	6.803 (1.84)	0.066	3.183 (0.99)	0.320
Log (Area size)	1.009 (0.03)	0.973	0.838 (−0.51)	0.608
Minority autonomous region	3.945 (1.65)	0.099	0.482 (−0.57)	0.568
Municipality directly under the central government	2.938 (1.82)	0.068	2.479 (1.15)	0.249
Year=2003	1.911 (1.07)	0.286	1.771 (0.91)	0.363
Year=2005	2.127 (0.99)	0.320	1.911 (0.85)	0.394
Year=2007	1.211 (0.21)	0.831	0.978 (−0.02)	0.982
Year=2009	7.136 (1.69)	0.092	5.564 (1.41)	0.159
Year=2011	4.688 (1.17)	0.240	3.520 (0.89)	0.373
Year=2013	1.431 (0.22)	0.824	0.984 (−0.01)	0.992
Year=2015	7.749 (1.21)	0.228	4.919 (0.87)	0.384
Cut1	3.88952e+11 (1.51)	0.132	2534148.2 (0.84)	0.401
Cut2	3.15425e+12 (1.62)	0.106	21532011.2 (0.95)	0.340
Observations	248		248	
Pseudo R-squared	0.181		0.190	

Note: Ordered logit models with two-tailed p values. The dependent variables are measured with a three-point scale, ranging from zero to two city-level IECLG winners or finalists in a province. Robust standard errors clustered by provinces.

comprising 2005, 2007, 2009, 2011, 2013, and 2015. The unit of analysis is city-year and the statistical analyses include around 1,700 observations.

The definition of all variables, along with detailed measurements and data sources are provided in Table 5.6. Our dependent variable, the county-level government innovativeness of a city, is measured by the number of county-level IECLG winners or finalists in a city in a year. Most of the cities in the sample had no IECLG winner or finalist in most years during the observed period, 49 city-years had one IECLG winner or finalist, and only

Table 5.6 Variables, measures, and data sources for city-level analysis (2005–2015)

Variables	Description of Measurement	Sources
Dependent variable		
Number of county-level IECLG winners or finalists	Number of county-level IECLG winners or finalists in a city in a year	The Websites of the Central Compilation and Translation Bureau, Party School of the CPC Central Committee, and Peking University.
Independent variable		
Number of county-level regions	The number of county-level regions (comprising urban districts and rural counties) in a city in a year	National Bureau of Statistics of China (www.data.stats.gov.cn)
Control variables		
Fiscal dependence (%)	The extent to which a city government depends on the higher-level governments for fiscal resources. This variable is calculated by $$\frac{\text{Expenditure}_{i,t-1} - \text{Revenue}_{i,t-1}}{\text{Expenditure}_{i,t-1}} \times 100$$ (Zhang 2015)	China Stock Market & Accounting Research Database (www.gtarsc.com)
Service sector (%)	Percentage of the service sector in the city's GDP in the previous year	
Log (Foreign investment)	Natural logarithm of contracted investment (ten thousand dollars) in the previous year	
Log (GDP)	Natural logarithm of GDP (100 million yuan) in a city in the previous year	
Log (Population)	Natural logarithm of the population size in a city in the previous year	
Log (Area size)	Natural logarithm of the area size (square kilometres) of a city	
Sub-provincial	Dummy=1 if a city is sub-provincial city (e.g., Changchun, Chengdu, Dalian, Guangzhou, Hangzhou, Harbin, Jinan, Nanjing, Ningbo, Qingdao, Shenyang, Shenzhen, Wuhan, Xi'an, and Xiamen) and zero otherwise	

60 An empirical investigation on the awards

one city-year (Ankang City in 2013) had two IECLG winners or finalists. In contract with the previous model choice in the provincial-level analysis, I employed logit models instead to conduct the statistical analyses since the dependent variable only has one observation with the value of two. I collected the information on the dependent variable from the websites of the CCTB, CPS, and PKU.

The key independent variable, the span of control in a city, is measured by the number of county-level administrative regions in a city in a year. I collected this data from the website of the NBS of China. According to the theoretical analyses in Chapter 3, I expected that the independent variable positively predicts the dependent variable.

I also included a number of variables to control for the characteristics of each city in our analysis. First, a control for fiscal dependence, which refers to the extent to which a city government depends on the superior governments for fiscal resources. This variable is calculated by $\frac{\text{Expenditure}_{i,t-1} - \text{Revenue}_{i,t-1}}{\text{Expenditure}_{i,t-1}} \times 100$ for city i in year t (Zhang 2015). Second, the percentage of the service sector in a city's GDP was included in the model to account for the potential effects of the economic structure. Third, I included the size of contracted investments (millions of dollars) in a city in a year in the models to account for the potential effects of economic openness. I took the natural logarithm of this variable to capture the potential relative change (e.g., a per cent change) rather than absolute change (e.g., a unit change) caused by the monetary variables with a highly skewed distribution. Fourth, similarly, I included the natural logarithm of GDP in a city in a year in the models to account for the potential effect of economic size on government innovativeness. Fifth, I included the natural logarithm of the population size in a city in a year in the models as a province with a bigger population size may have more diverse demands for public service innovations.

Sixth, I included the natural logarithm of the area size (square kilometres) of a city in the models because the area size of a city may determine both the number of county-level regions in a city (i.e., the independent variable) and the diversity or intensity of public demands for government innovativeness (i.e., the dependent variable). Seventh, I further controlled for the administrative categories of each city-level unit, comprising normal prefecture-level cities and sub-provincial cities. Specifically, I created a dummy equal to one if the observed unit is a sub-provincial city, such as Changchun, Chengdu, Dalian, Guangzhou, Hangzhou, Harbin, Jinan, Nanjing, Ningbo, Qingdao, Shenyang, Shenzhen, Wuhan, Xi'an, and Xiamen; and zero otherwise. In other words, the reference category was the normal prefecture-level cities.

Finally, in the statistical models, I lagged the time-varying city-level control variables by one year to avoid reverse causation (Zhu and Zhang 2016, 2019). My analysis also accounted for the potential temporal dependence problem by adding the year fixed effects.

Table 5.7 reports the summary statistics of all variables in the city-level statistical analysis (2005–2015). On average, a city had 0.03 city-level IECLG winners or finalists in a year; the minimum value is zero and the maximum value is two. Moreover, on average, the number of county-level regions in a city has a mean value of 8.24 and ranges from 0 (i.e., Dongguan, Zhongshan, Sanya, Sansha, and Jiayuguan in certain years during the observation period) to 25 (i.e., Baoding). The fiscal dependence on the superior governments has an average value of 51.06 and ranges from –25.58 to 74.85. The percentage of the service sector in the GDP of a province has an average value of 35.69 and ranges from 8.59 to 74.85. The natural logarithm of foreign investments (ten thousand dollars) has an average value of 9.49 and ranges from 0.69 to 14.15. The natural logarithm of GDP (100 million yuan) has an average value of 6.62 and ranges from 3.58 to 9.72. The natural logarithm of population size has an average value of 15.03 and ranges from 6.91 to 16.33. The natural logarithm of area size (square kilometres) has an average value of 9.33 and ranges from 2.56 to 12.44. A total of 4 per cent of the observations in the sample belong to the sub-provincial cities.

Table 5.8 reports the estimates of county-level IECLG winners or finalists in a city. Specifically, Table 5.8 present specifications with controls only

Table 5.7 Summary statistics for city-level analysis (2005-2015)

Variable	Observations	Mean	Standard. Deviation	Minimum Value	Maximum Value
Number of county-level IECLG winners or finalists	1,699	0.03	0.18	0.00	2.00
Number of county-level regions	1,699	8.24	3.82	0.00	25.00
Fiscal dependence (%)	1,694	51.06	22.80	-25.58	93.71
Service sector (%)	1,694	35.69	8.23	8.58	74.85
Log (Foreign investment)	1,610	9.49	1.88	0.69	14.15
Log (GDP)	1,694	6.62	1.00	3.58	9.72
Log (Population)	1,695	15.03	0.70	6.91	16.33
Log (Area size)	1,699	9.33	0.85	2.56	12.44
Sub-provincial city	1,699	0.04	0.21	0.00	1.00

Table 5.8 City-level analysis results (2005–2015)

	Model 3		Model 4	
	Odds Ratio (t-statistic)	p-value	Odds Ratio (t-statistic)	p-value
Number of county-level regions			1.101 (1.95)	0.051
Fiscal dependence (%)	0.998 (−0.17)	0.864	0.997 (−0.23)	0.820
Service sector (%)	1.002 (0.08)	0.940	1.000 (0.01)	0.995
Log (Foreign investment)	0.912 (−0.56)	0.575	0.932 (−0.43)	0.668
Log (GDP)	4.276 (2.58)	0.010	4.066 (2.53)	0.011
Log (Population)	0.547 (−1.33)	0.183	0.378 (−1.86)	0.063
Log (Area size)	1.118 (0.54)	0.586	0.941 (−0.27)	0.786
Sub-provincial city	1.644 (1.03)	0.301	1.477 (0.82)	0.412
Year=2007	0.458 (−1.56)	0.119	0.468 (−1.52)	0.127
Year=2009	0.319 (−2.13)	0.033	0.334 (−2.07)	0.038
Year=2011	0.164 (−2.51)	0.012	0.175 (−2.47)	0.014
Year=2013	0.111 (−2.55)	0.011	0.121 (−2.52)	0.012
Year=2015	0.117 (−2.50)	0.012	0.128 (−2.47)	0.014
Constant	0.0336 (−0.74)	0.462	22.30 (0.54)	0.586
Observations	1610		1610	
Pseudo R-squared	0.147		0.153	

Note: Logit models with two-tailed *p* values. The dependent variables are measured with a two-point scale, ranging from zero to one county-level IECLG winners or finalists in a city. Robust standard errors clustered by cities.

alongside specifications that include the key independent variable, the number of county-level regions in a city. The key independent variable improves the Pseudo R-squared by 0.06, indicating that the number of county-level IECLG winners or finalists in a city varies significantly by the number of county-level regions in a city. The odds ratio for one unit increase in each independent variable and *t*-statistics in parentheses are provided. The two-tailed *p* values are shown in separate columns.

The results are consistent with previous hypotheses regarding the relationship between the span of control and local government innovativeness. Specifically, as Model 4 shows, one more county-level region in a city is associated with a 10.1 per cent increase in the odds of having one more county-level IECLG winner or finalist in a city in a year, with a *p*-value of 0.051. According to the theoretical analysis in Chapter 3, this evidence suggests that the number of county-level regions in a city may be positively related to county governments' policy autonomy and county leaders' competition intensity and horizontal transfer opportunities, thus increasing the overall county-level government innovativeness in a city.

Several city-level characteristics yield notable results, too. City-level GDP is significantly and positively correlated with the number of county-level IECLG winners or finalists, indicating that economic development might provide the necessary resources for adopting innovations. City-level population size is significantly and negatively correlated with the number of county-level IECLG winners or finalists. One potential explanation for this finding is that local governments that serve a large population may face more obstacles when introducing innovations in local public management; they may also invest more resources in providing basic public services and allocate fewer resources for new policies or programs given the potential costs or risks associated with possible failures. Moreover, the statistically significant coefficients of the year dummies show that there have been generally fewer county-level innovations in recent years, which is consistent with other scholars' observations about the notable reduction in local policy experimentation in China due to the decreased local policy autonomy under President Xi Jinping's rule (Hasmath, Teets, and Lewis 2019; Teets and Hasmath 2020). Nevertheless, we do not have strong evidence to support the potential effects of other city-level socioeconomic or geographic characteristics on county-level government innovativeness.

Bibliography

Almén, Oscar. "Local participatory innovations and experts as political entrepreneurs: The case of China's democracy consultants." *Democratization* 23, no. 3 (2016): 478–497.

Bernier, Luc, Taïeb Hafsi, and Carl Deschamps. "Environmental determinants of public sector innovation: A study of innovation awards in Canada." *Public Management Review* 17, no. 6 (2015): 834–856.

Berry, Frances Stokes, and William D. Berry. "State lottery adoptions as policy innovations: An event history analysis." *American Political Science Review* 84, no. 2 (1990): 395–415.

Chung, Jae Ho, and Tao-chiu Lam. "China's "city system" in flux: Explaining postmao administrative changes." *The China Quarterly* 180 (2004): 945–964.

Hasmath, Reza, Jessica C. Teets, and Orion A. Lewis. "The innovative personality? Policy making and experimentation in an authoritarian bureaucracy." *Public Administration and Development* 39, no. 3 (2019): 154–162.

Huang, Bin, Mengmeng Gao, Caiqun Xu, and Yu Zhu. "The impact of province-managing-county fiscal reform on primary education in China." *China Economic Review* 45 (2017): 45–61.

Ma, Liang. "Political ideology, social capital, and government innovativeness: Evidence from the US states." *Public Management Review* 19, no. 2 (2017): 114–133.

Mohr, Lawrence B. "Determinants of innovation in organizations." *American Political Science Review* 63, no. 1 (1969): 111–126.

Teets, Jessica C., and Reza Hasmath. "The evolution of policy experimentation in China." *Journal of Asian Public Policy* 13, no. 1 (2020): 49–59.

Walker, Richard M., Frances S. Berry, and Claudia N. Avellaneda. "Limits on innovativeness in local government: Examining capacity, complexity, and dynamism in organizational task environments." *Public Administration* 93, no. 3 (2015): 663–683.

Wang, Qinghua, and Gang Guo. "Yu Keping and Chinese intellectual discourse on good governance." *The China Quarterly* 224 (2015): 985–1005.

Wu, Jiannan, Liang Ma, and Yuqian Yang. "Innovation in the Chinese public sector: Typology and distribution." *Public Administration* 91, no. 2 (2013): 347–365.

Zhang, Yanlong. "The formation of public-private partnerships in China: An institutional perspective." *Journal of Public Policy* 35, no. 2 (2015): 329.

Zhao, Qiang. "The regional disparities in Chinese provincial government innovation." *Innovation* 14, no. 4 (2012): 595–604.

Zhu, Xufeng. "Mandate versus championship: Vertical government intervention and diffusion of innovation in public services in authoritarian China." *Public Management Review* 16 (2014): 117–139.

Zhu, Xufeng, and Youlang Zhang. "Political mobility and dynamic diffusion of innovation: The spread of municipal pro-business administrative reform in China." *Journal of Public Administration Research and Theory* 26, no. 3 (2016): 535–551.

Zhu, Xufeng, and Youlang Zhang. "Diffusion of marketization innovation with administrative centralization in a multilevel system: Evidence from China." *Journal of Public Administration Research and Theory* 29, no. 1 (2019): 133–150.

6 Conclusion

In recent decades, various types of innovation in the public sector, such as performance management, privatisation, public–private partnership, outsourcing, and e-government have become increasingly popular in both developed and developing countries (Ansell and Gash 2008; Walker, Damanpour, and Devece 2011). Since the 1960s, the vast amount of literature on local government innovation has been seeking to explain the origins, strategies, content, and consequences of specific innovation cases (Mohr 1969; Berry and Berry 1990; Shipan and Volden 2008). However, the organisational innovativeness of local governments is often ignored. Given the fact that the extent and frequency of adopting innovation varies greatly across local governments around the world (Ma 2017), a systematic analysis of local government innovativeness is theoretically, normatively, and practically significant. Moreover, it was not until the 2010s that scholars systematically attempted to use empirical data to examine local government innovations in non-Western contexts, such as China. More government innovation studies in non-Western contexts are highly encouraged (Walker, Avellaneda, and Berry 2011).

To improve our understanding of local government innovativeness in China, this study adopted an institutional approach to create a novel theoretical framework to explain why some governments in China are more innovative than others. This study first analysed existing explanations for local government innovation in China, comprising internal factors (e.g., the macro-level jurisdictional characteristics and government characteristics, and the micro-level policymaker characteristics and policy entrepreneurship) and external pressures (e.g., the vertical top-down and bottom-up mechanisms and the horizontal learning, imitation, and competition mechanisms). This book further highlights how little attention has been paid to the structural characteristics of China's government organisations, such as the administrative hierarchy and span of control.

66 Conclusion

The theoretical framework proposed in this book builds on insights from the institutional analysis of decentralisation (Bardhan 2002; Faguet 2014; Qian, Roland, and Xu 2006), promotion tournament (Li and Zhou 2005; Lü and Landry 2014), and career mobility (Opper, Nee, and Brehm 2015; Walder 1995; Zhu 2018) in China's hierarchical government system in previous literature. This book indicates that the positions of local governments in the administrative hierarchy are positively associated with their innovation resources and negatively associated with their innovation motivations. Since both motivations and resources are necessary but insufficient conditions for generating innovations, this book proposes an inverse U-shaped relationship between the administrative ladder and local government innovativeness in China and argues that other things being equal, compared to the high-level (e.g., provincial-level) or low-level (e.g., township-level) local governments, the middle-level governments are more innovative given fewer constraints on their innovation resources and innovation motivations.

In addition, this book shows that the span of control could increase local government innovativeness in three ways. First, based on the decentralisation literature, as the span of control increases, the policy autonomy of local governments tends to increase, which could further increase the capability of local governments to innovate. Second, based on the promotion tournament literature, as the span of control increases, the competition intensity between local leaders increases, which further increases their motivation to innovate to outperform their peers. Third, based on the career mobility literature, as the span of control increases, the opportunities for horizontal transfer or rotation tend to increase, which could further increase the opportunities of local leaders to innovate.

This study employed empirical data drawn from multiple sources to demonstrate support for the above theoretical expectations. However, before the statistical analysis, this study first conducted a detailed review of local governments' innovative practices in China by examining the IECLG Awards Programme from 2001 to 2015. Local governments were encouraged to submit application materials to the IECLG team every two years. The descriptive analysis suggests that the coastal local governments are generally more innovative than the inland local governments. Local governments have actively adopted management innovations and service innovations to improve business environments and attract investments in the past two decades.

Further statistical analysis provides supportive evidence for the theoretical expectations of administrative hierarchy and the span of control. For instance, among the 177 winners and finalists in the eight waves of IECLG competition, 147 belonged to city-level governments and the county-level governments. Moreover, a series of provincial-level and city-level panel data analyses show that a province's number of city-level regions is significantly positively associated with that province's number of city-level IECLG winners or finalists, and a city's number of county-level regions is significantly positively associated with that city's number of county-level IECLG winners or finalists.

This book makes multiple contributions. First, in contrast to previous explanations for local government innovations, based on internal determinants or external pressures, this book provides the first attempt to establish a theoretical logic that associates the institutional characteristics of China's multi-level government structure with local government innovativeness. The exploration of the institutional factors underlying local government innovativeness could help understand why some local governments are systematically more innovative than others. Moreover, previous studies on local government innovation in China tended to focus on certain specific innovations, this study took a step back and explored the topic from an organisational perspective, generating findings that are widely applicable in the country's system of governance. Following the approach adopted by the most recent government innovation research based on a well-established database that consists of hundreds of innovation cases (Boehmke et al. 2020), this study paved the way for rigorous statistical analysis of the organisational innovativeness of local governments in China in view of the inherent limits of most of the studies based on only one or several innovation cases.

Second, this book contributes to the research on subnational intergovernmental relations. Most of the extant studies on intergovernmental relations tend to focus on the interactions between the national and subnational governments and potentially ignore the vertical and horizontal interactions between subnational governments (Cai and Treisman 2005; Qian and Weingast 1997; Schneider 2003; Xu 2011). This book points out that the administrative hierarchy and the span of control can shape how local governments interact with each other and choose to innovate or not. Given the potentially significant socioeconomic consequences created by the variation of local government innovativeness, my exploration into the effects of administrative hierarchy and the span of control on local government innovativeness can provide the scholarly foundations for future efforts of examining how subnational intergovernmental relations shape socioeconomic development in a country.

Third, this work has also contributed to the study of organisational management in the public sector. It advances the literature by showing that government structures could significantly shape local government decisions and may have broader implications regarding the local resource allocation process. Moreover, public management scholars tend to view organisational innovation as an important determinant of public organisation performance (Damanpour and Evan 1984; Walker, Damanpour, and Devece 2011). However, only a limited number of studies have empirically studied the origins of public organisational innovativeness (Ma 2017; Walker, Berry, and Avellaneda 2015). As local government innovation awards programmes (e.g., IECLG in China or IAGA in the United States) provide an objective and comparable measure of government innovativeness, this book further sheds some new light on the explanations for organisational innovativeness in public management research.

Fourth, the patterns of local government innovativeness in China identified in this book could potentially be generalised in other contexts. In modern countries with a multiple-level power structure, the distribution of various types of resources and incentives along government hierarchy and the span of control tends to be imbalanced. Under these circumstances, the motivations and capabilities of local governments for adopting innovations are significantly shaped by their positions in the government hierarchy and the number of competing peers. Studying the dynamics underlying these processes could improve the academic knowledge of local governments' decisions and behaviours and help practitioners to use this knowledge to promote or revise certain designs of government structures and address socioeconomic injustices produced by the unequal distribution of government innovativeness.

Fifth, practically, this book provides important insights to government reforms in China. In recent years, China has initiated a series of reforms regarding organisational structures in the public sector, such as the "Province-Managing-County Reform", the "Turning the Prefectures into Cities Reform", the "Super-Department Reform", and the "Multiple-Plan Integration Reform". In response to these administrative reforms, this book provides theoretical insights and empirical evidence to support the restructuring of the political structure, the flattening the administrative hierarchy, and the decreasing of monitoring costs to make it easier for local governments to act as policy entrepreneurs and adopt innovative practices to improve public services. Policymakers should identify the scope and depth of government innovativeness shaped by administrative hierarchy and the span of control before they decide to promote innovations among local governments.

Moreover, this work suggests that the hierarchical bureaucracy in China provides an important institutional basis for local government innovations. However, a recent study by Teets and Hasmath (2020) found that President Xi Jinping's centralised policies under the guise of "top-level design" and his anticorruption campaign have created significant disincentives for local government innovations. In spite of this, Chinese local governments have not stopped being innovative. For instance, during the recent COVID-19 pandemic, the Zhejiang province initiated multiple emergency management innovations, such as using the strengths of community-based organisations in multiple stages of COVID-19 responses and creating digital tracking platforms (Cheng et al. 2020). The theoretical analysis of administrative hierarchy and span of control in this book suggests that there are multiple institutional incentives embedded in China's hierarchical bureaucracy that encourage local governments to be innovative.

There are multiple possible directions for future research. First, although Chapter 3 has highlighted several theoretically plausible and practically intuitive mechanisms underlying the main theoretical expectations, some cannot be directly empirically verified given the current data or resource limit. For example, due to the lack of systematic elite surveys or interviews across the administrative hierarchy, it is difficult to directly verify whether government leaders have weaker innovation motivations when they are promoted to higher positions. It is also empirically difficult to directly quantify and compare the resources allocated for innovative practices by each level of government.

Moreover, there is not enough information to systematically analyse how the span of control affects political elites' perceptions of policy autonomy, competition intensity, and horizontal transfer opportunities and how these factors affect their specific choices in terms of adopting or implementing innovative practices. Admittedly, understanding how government leaders make decisions in practice can help complement the quantitative analysis by adding traction to the causality underlying the statistical analyses. Due to the unexpected COVID-19 pandemic, my research plan in terms of conducting further field observations and qualitative interviews has been largely disrupted. Researchers with more resources or funding may build on this book to further verify or develop these theoretical mechanisms.

Also, we still know little about how local government innovativeness shapes the development of public organisations, public opinions, and social or economic outcomes. Theoretically, according to the policy feedback theory (Moynihan and Soss 2014; Nowlin 2016; Pierson 1993, 2000), local government innovativeness can be political forces in their own right that can affect organisational capacity, organisational structure, working routines, bureaucratic authorities, public motivations, cultures, and public perceptions of governments. The innovativeness of a local government may also affect how local interest groups are mobilised in policy subsystems and the frequency or intensity of policy changes, thus shaping local socioeconomic development. Therefore, future research needs to study how, when, and where local government innovativeness produce certain consequences.

Conclusion

Additionally, the organisational innovativeness and specific innovations of governments are not always beneficial. Government innovations are not only associated with risks and costs for governments themselves, but also pose potential threats to civil rights. For instance, the so-called "grid governance (GG)" in China's urban neighbourhoods, first initiated in Zhejiang province and then widely implemented in China in recent years, has become an important policy tool for grassroots mobilisation and conflict resolution to the detriment of citizens' convenience, freedom, and privacy (Tang 2020). Therefore, the potential negative outcomes of local government innovativeness and specific innovations deserve more attention in future research.

Finally, comparative studies of government innovativeness with cases from more countries can help produce insights that are otherwise overlooked. The literature review presented in this book shows that only a limited number of studies have explored the determinants of local government innovativeness in a few countries, such as the United Kingdom, Canada, and the United States. Nevertheless, as this book shows, a detailed analysis of a typically non-Western country, China, could provide important theoretical insights into how the institutional characteristics of the government structure can shape organisational innovativeness in the public sector. Future research may further examine whether the hypotheses generated based on the case of China can be generalised in other contexts and whether the effects of institutions vary under different circumstances.

Bibliography

Ansell, Chris, and Alison Gash. "Collaborative governance in theory and practice." *Journal of Public Administration Research and Theory* 18, no. 4 (2008): 543–571.

Bardhan, Pranab. "Decentralization of governance and development." *Journal of Economic Perspectives* 16, no. 4 (2002): 185–205.

Berry, Frances Stokes, and William D. Berry. "State lottery adoptions as policy innovations: An event history analysis." *American Political Science Review* 84, no. 2 (1990): 395–415.

Boehmke, Frederick J., Mark Brockway, Bruce A. Desmarais, Jeffrey J. Harden, Scott LaCombe, Fridolin Linder, and Hanna Wallach. "SPID: A new database for inferring public policy innovativeness and diffusion networks." *Policy Studies Journal* 48, no. 2 (2020): 517–545.

Cai, Hongbin, and Daniel Treisman. "Does competition for capital discipline governments? Decentralization, globalization, and public policy." *American Economic Review* 95, no. 3 (2005): 817–830.

Cheng, Yuan, Jianxing Yu, Yongdong Shen, and Biao Huang. "Coproducing responses to COVID-19 with community-based organizations: Lessons from Zhejiang Province, China." *Public Administration Review* 80, no. 5 (2020): 866–873.

Damanpour, Fariborz, and William M. Evan. "Organizational innovation and performance: The problem of" organizational lag." *Administrative Science Quarterly* 29, no. 3 (1984): 392–409.

Faguet, Jean-Paul. "Decentralization and governance." *World Development* 53 (2014): 2–13.

Li, Hongbin, and Li-An Zhou. "Political turnover and economic performance: The incentive role of personnel control in China." *Journal of Public Economics* 89, no. 9–10 (2005): 1743–1762.

Lü, Xiaobo, and Pierre F. Landry. "Show me the money: Interjurisdiction political competition and fiscal extraction in China." *American Political Science Review* 108, no. 3 (2014): 706–722.

Ma, Liang. "Political ideology, social capital, and government innovativeness: Evidence from the US states." *Public Management Review* 19, no. 2 (2017): 114–133.

Mohr, Lawrence B. "Determinants of innovation in organizations." *American Political Science Review* 63, no. 1 (1969): 111–126.

Moynihan, Donald P., and Joe Soss. "Policy feedback and the politics of administration." *Public Administration Review* 74, no. 3 (2014): 320–332.

Nowlin, Matthew C. "Policy change, policy feedback, and interest mobilization: The politics of nuclear waste management." *Review of Policy Research* 33, no. 1 (2016): 51–70.

Opper, Sonja, Victor Nee, and Stefan Brehm. "Homophily in the career mobility of China's political elite." *Social Science Research* 54 (2015): 332–352.

Pierson, Paul. "When effect becomes cause: Policy feedback and political change." *World Politics* 45, no. 4 (1993): 595–628.

Pierson, Paul. "Increasing returns, path dependence, and the study of politics." *American Political Science Review* 91, no. 2 (2000): 251–267.

Qian, Yingyi, Gerard Roland, and Chenggang Xu. "Coordination and experimentation in M-form and U-form organizations." *Journal of Political Economy* 114, no. 2 (2006): 366–402.

Qian, Yingyi, and Barry R. Weingast. "Federalism as a commitment to reserving market incentives." *Journal of Economic Perspectives* 11, no. 4 (1997): 83–92.

Schneider, Aaron. "Decentralization: Conceptualization and measurement." *Studies in Comparative International Development* 38, no. 3 (2003): 32–56.

Shipan, Charles R., and Craig Volden. "The mechanisms of policy diffusion." *American Journal of Political Science* 52, no. 4 (2008): 840–857.

Tang, Beibei. "Grid governance in China's urban middle-class neighbourhoods." *The China Quarterly* 241 (2020): 43–61.

Teets, Jessica C., and Reza Hasmath. "The evolution of policy experimentation in China." *Journal of Asian Public Policy* 13, no. 1 (2020): 49–59.

Walder, Andrew G. "Career mobility and the communist political order." *American Sociological Review* 60, no. 3 (1995): 309–328.

Walker, Richard M., Claudia N. Avellaneda, and Frances S. Berry. "Exploring the diffusion of innovation among high and low innovative localities: A test of the Berry and Berry model." *Public Management Review* 13, no. 1 (2011): 95–125.

Conclusion

Walker, Richard M., Frances S. Berry, and Claudia N. Avellaneda. "Limits on innovativeness in local government: Examining capacity, complexity, and dynamism in organizational task environments." *Public Administration* 93, no. 3 (2015): 663–683.

Walker, Richard M., Fariborz Damanpour, and Carlos A. Devece. "Management innovation and organizational performance: The mediating effect of performance management." *Journal of Public Administration Research and Theory* 21, no. 2 (2011): 367–386.

Xu, Chenggang. "The fundamental institutions of China's reforms and development." *Journal of Economic Literature* 49, no. 4 (2011): 1076–1151.

Zhu, Xufeng. "Executive entrepreneurship, career mobility and the transfer of policy paradigms." *Journal of Comparative Policy Analysis: Research and Practice* 20, no. 4 (2018): 354–369.

Appendix

Table A.1 A list of all winners and finalists in the IECLG Awards Programme (2001–2015)

Year	Location	Initiator	Innovation	Award type	Innovation typology
2001	Shizhong District, Suining City, Sichuan Province	District government	Public recommendation and the democratic election of township party chief and government chief	Winner	Governance innovation
2001	Qianxi County, Tangshan City, Hebei Province	County government	Direct election of women's congress	Winner	Governance innovation
2001	Nanning City, Guangxi Zhuang Autonomous Region	City government	The promotion of the government procurement system	Winner	Management innovation
2001	Xiaguan District, Nanjing City, Jiangsu Province	District government	The creation of "Government Supermarket"	Winner	Service innovation
2001	Jinhua City, Zhejiang Province	City government	The auditing of government officials' economic responsibility	Winner	Governance innovation
2001	Guiyang City, Guizhou Province	City-level people's congress standing committee	The promotion of civic hearings	Winner	Governance innovation
2001	Shenzhen City, Guangdong Province	City government	Reform of the administrative examination and approval system	Winner	Management innovation
2001	Pudong New Area, Shanghai City	District government	Creation of the social conflict mediation centre	Winner	Management innovation
2001	Haikou City, Hainan Province	City government	The "three systems" for the reform of administrative examination and approval system	Winner	Management innovation

2001	Guangshui City, Suizhou City, Hubei Province	City government	"Two-vote system" in the election of the village party chief	Winner	Governance innovation
2003	Ganchahe Town, Shucheng County, Anhui Province	Township government	Privatisation of nonprofit undertakings in small towns	Winner	Collaborative innovation
2003	Shenzhen City, Guangdong Province	City government	Public utility marketisation reform	Winner	Collaborative innovation
2003	Nanning City, Guangxi Zhuang Autonomous Region	City government	Social emergency collaboration system	Winner	Collaborative innovation
2003	Shijiazhuang City, Hebei Province	City government	Children protection education centre	Winner	Service innovation
2003	Longhua District, Haikou City, Hainan Province	District government	Home for migrant workers	Winner	Service innovation
2003	Lishu County, Siping City, Jilin Province	County government	Public election of village committee	Winner	Governance innovation
2003	Buyun Township, Shizhong District, Suining City, Sichuan Province	Township government	Direct election of village chief	Winner	Governance innovation
2003	Qingdao City, Shandong Province	City government	"Sunshine Rescue" project	Winner	Service innovation
2003	Huzhou City, Zhejiang Province	City government	Reform of household registration system	Winner	Service innovation
2003	Wenling City, Taizhou City, Zhejiang Province	City government	Democratic deliberation	Winner	Governance innovation
2005	Yantian District, Shenzhen City, Guangdong Province	District government	Community management system reform	Winner	Management innovation

(*Continued*)

76 Appendix

Table A.1 Continued

Year	Location	Initiator	Innovation	Award type	Innovation typology
2005	Pingchang County, Bazhong City, Sichuan Province	County government	Public recommendation and direct election of township party committee members	Winner	Governance innovation
2005	Chongqing Municipality	Provincial-level government	Four systems of creating "government by law"	Winner	Governance innovation
2005	Quanzhou City, Fujian Province	City-level federation of trade unions	New model for the protection of the rights of migrant workers	Winner	Service innovation
2005	Qian'an City, Tangshan City, Hebei Province	City government	New rural cooperative medical system	Winner	Service innovation
2005	Guangxi Zhuang Autonomous Region	Provincial-level department of civil affairs	Construction of the "Five guarantee villages"	Winner	Service innovation
2005	Hunan Province	Provincial-level women's federation	Rural women' participation in village-level governance	Winner	Governance innovation
2005	Shijingshan District, Beijing Municipality	District government and CCP committee	Lugu community street management system innovation	Winner	Management innovation
2005	Jialian Street Office, Siming District, Xiamen City, Fujian Province	Township-level office of street affairs	"Loving Care Supermarket"	Winner	Service innovation
2005	Nankai District, Tianjin City	District-level administrative licensing centre	"Overtime acquiescence" mechanism	Winner	Service innovation
2007	Yiwu City, Jinhua City, Zhejiang Province	City-level federation of trade unions	The socialisation of right protection by trade union	Winner	Service innovation

Appendix 77

2007	Haishu District, Ningbo City, Zhejiang Province	District government	Government-paid home care services	Winner	Service innovation
2007	Yichun City, Heilongjiang Province	City government	Reform of forestry property right system	Winner	Management innovation
2007	Nanshan District, Shenzhen City, Guangdong Province	District government	Two-way interaction system for harmonious community construction	Winner	Collaborative innovation
2007	Rushan City, Weihai City, Shandong Province	City-level CCP committee	Promoting democracy within the Party	Winner	Governance innovation
2007	Changshou Road, Putuo District, Shanghai City	Township-level office of street affairs	Community civil organisation management system reform	Winner	Management innovation
2007	Xian'an District, Xianning City, Hubei Province	District-level CCP committee	Reform of township administrative management system	Winner	Management innovation
2007	Jiangsu Province	Provincial-level department of public security	Law enforcement notification service system	Winner	Service innovation
2007	Chengdu City, Sichuan Province	City government	Deepening the reform of administrative examination and approval system	Winner	Management innovation
2007	Laixi City, Qingdao City, Shandong Province	City government	Public service agent system	Winner	Service innovation
2009	Inner Mongolia	Provincial-level public security frontier corps	"Prairie 110"	Winner	Service innovation
2009	Qingdao City, Shandong Province	City government and CCP committee	Diversified official accountability system	Winner	Governance innovation
2009	Shiquan County, Ankang City, Shaanxi Province	County government and CCP committee	Constructing a long-term mechanism for caring for left-behind children	Winner	Service innovation

(*Continued*)

Table A.1 Continued

Year	Location	Initiator	Innovation	Award type	Innovation typology
2009	Hangzhou City, Zhejiang Province	City government	Open decision-making	Winner	Governance innovation
2009	Shenzhen City, Guangdong Province	City-level bureau of social organisation management	Reform of social organisation registration and management system	Winner	Service innovation
2009	Beijing Municipality	Provincial-level government	"Three Effects and One Innovation" performance management system	Winner	Management innovation
2009	Xiamen City, Fujian Province	City government	Construction of citizen health information system	Winner	Technological innovation
2009	Kuitun Tianbei New District, Seventh Agricultural Division of Xinjiang Corps	County government	Military and Local public integration management system	Winner	Collaborative innovation
2009	Shenyang City, Liaoning Province	City government and CCP committee	New mechanism for petitions and visits	Winner	Service innovation
2009	Jiangyin City, Wuxi City, Jiangsu Province	City government and CCP committee	Construction of the comprehensive evaluation index System of "Happy Jiangyin"	Winner	Technological innovation
2011	Guangdong Province	Provincial-level office of organisation and bianzhi committee	The super-department system reform	Winner	Management innovation
2011	Zichang County, Yan'an City, Shaanxi Province	County government	Public hospital reform	Winner	Management innovation
2011	Pudong New District, Shanghai Province	District-level bureau of civil affairs	Public service park	Winner	Service innovation

Year	Location	Organization	Innovation	Result	Type
2011	Cixi City, Ningbo City, Zhejiang Province	City government and CCP committee	Cooperative governance model of grassroots government organisations and social organisations	Winner	Collaborative innovation
2011	Hebei Province	Provincial-level department of environmental protection	Watershed ecological compensation mechanism	Winner	Collaborative innovation
2011	Liaoning Province	Provincial-level discipline inspection commission, government supervision department, and government correction office	"Minxin Network"	Winner	Technological innovation
2011	Wanzai County, Yichun City, Jiangxi Province	County government	A new model of the localisation of rural social work	Winner	Management innovation
2011	Hainan Province	Provincial-level government service centre	"Three Concentrations" of administrative examination and approval affairs and powers	Winner	Management innovation
2011	Suining City, Sichuan Province	City-level political and legal committee	The social stability risk assessment mechanism for major events	Winner	Management innovation
2011	Shaoxing City, Zhejiang Province	City government	Power regulation of the central towns	Winner	Management innovation
2013	Jiangxi Province	Provincial-level department of justice	Innovating the resettlement model	Winner	Service innovation

(*Continued*)

80 Appendix

Table A.1 Continued

Year	Location	Initiator	Innovation	Award type	Innovation typology
2013	Guizhou Province	Provincial committee of the CYLC's Chunhui action development centre	Chunhui Action	Winner	Service innovation
2013	Zhangpu Town, Kunshan City, Jiangsu Province	Township government and CCP committee	Administrative reform and government process reengineering in economically developed towns	Winner	Management innovation
2013	Sichuan Province	Provincial-level disabled persons' federation	Tailored service mode for disabled people	Winner	Service innovation
2013	Zhongshan City, Guangdong Province	City-level social work committee	The point-based migrant management system	Winner	Management innovation
2013	Chengdu City, Sichuan Province	City government	Reform of rural property rights system	Winner	Management innovation
2013	Antu County, Yanbian Korean Autonomous Prefecture, Jilin Province	County government and CCP committee	The creation of people's appeal service platform	Winner	Technological innovation
2013	Jiaozuo City, Henan Province	City-level bureau of finance	"Separation of four powers" in new financial management system	Winner	Management innovation
2013	Langao County, Ankang City, Shaanxi Province	County government	Reform of new rural cooperative reimbursement system in town-run health centres	Winner	Management innovation

Appendix 81

2013	Shangcheng District, Hangzhou City, Zhejiang Province	District government and CCP committee	The standardisation of government management and public services	Winner	Service innovation
2015	Qingdao City, Shandong Province	City-level bureau of human resources and social insurance	Long-term medical insurance system	Winner	Management innovation
2015	Yantian District, Shenzhen City, Guangdong Province	District government	Municipal ecological gross product value accounting system and its application	Winner	Service innovation
2015	Shunde District, Foshan City, Guangdong Province:	District government	Public decision advisory committee system	Winner	Governance innovation
2015	Nanling County, Wuhu City, Anhui Province	County government and CCP committee	"Three types of meetings, four types of self-governance, and one platform" governance model for public construction in rural areas	Winner	Governance innovation
2015	Xicheng District, Beijing Municipality	District government	Fully Responsive grid-based social service model	Winner	Governance innovation
2015	Dadukou District, Chongqing Municipality	District-level people's procuratorate	"Sister Sha" crime prevention and education system	Winner	Service innovation
2015	Wuchang District, Wuhan City, Hubei Province	District government	Using nonprofit enterprises to promote social governance innovation	Winner	Governance innovation
2015	Xinzhou City, Shanxi Province	City-level publicity department of the CCP committee	Xinzhou Snapshot	Winner	Technological innovation

(*Continued*)

Appendix

Table A.1 Continued

Year	Location	Initiator	Innovation	Award type	Innovation typology
2015	Zhejiang Province	Provincial-level office of organisation and bianzhi committee	Streamlining government functions and powers, and promotes the list of powers system	Winner	Governance innovation
2015	Changsha City, Hunan Province	City-level legislative affairs office	The standardisation and formalisation of the government's legal system	Winner	Governance innovation
2001	Dapeng Town, Shenzhen City, Guangdong Province	Township government	"Two votes in three rounds" for electing the township chief	Finalist	Governance innovation
2001	Jinping County, Honghe Hani and Yi Autonomous Prefecture, Yunnan Province	County government	Poverty alleviation project	Finalist	Service innovation
2001	Quzhou City, Zhejiang Province	City government	Agricultural technology 110	Finalist	Technological innovation
2001	Shuyang County, Suqian City, Jiangsu Province	County government	Public notice before the formal appointment of cadres	Finalist	Governance innovation
2001	Pingchang County, Bazhong City, Sichuan Province	County government	Public tax assessment system	Finalist	Governance innovation
2001	Kangjian Street, Xuhui District, Shanghai City	Township-level office of street affairs	Recreation project	Finalist	Service innovation
2001	Qidaowan Township, Urumqi City, Xinjiang Uygur Autonomous Region	Township government	Transparent village affairs	Finalist	Governance innovation

Appendix 83

2001	Hefeng County, Enshi Tujia and Miao Autonomous Prefecture, Hubei Province	County government	Private business owner responsibility system for poverty alleviation projects	Finalist	Collaborative innovation
2001	Sheqi County, Nanyang City, Henan Province	County government	Top-down investigation group	Finalist	Governance innovation
2001	Changsha City, Hunan Province	City government	Four-level combined transparency of government affairs	Finalist	Governance innovation
2003	Ya'an City, Sichuan Province	City government	Direct election of county-level party representatives	Finalist	Governance innovation
2003	Taizhou City, Zhejiang Province	City government	Direct election of the chief of the township-level (or street-level) CYLC committee	Finalist	Governance innovation
2003	Nanning City, Guangxi Zhuang Autonomous Region	City government	Reform of state-owned assets management system	Finalist	Management innovation
2003	Yanqing County, Beijing	County government	Initiative of "stop and prevent domestic violence"	Finalist	Service innovation
2003	Siming District, Xiamen City, Fujian Province	District government	Public sector performance evaluation	Finalist	Management innovation
2003	Jiaozuo City, Henan Province	City government	The construction of "three-level" service-oriented government	Finalist	Service innovation
2003	Qianxi County, Tangshan City, Hebei Province	County government	Women rights protection	Finalist	Service innovation
2003	Beijing Municipality	Provincial-level government	Community public service platform	Finalist	Service innovation
2005	Shenhe District, Shenyang City, Liaoning Province	District government	Honesty system construction	Finalist	Management innovation

(*Continued*)

Table A.1 Continued

Year	Location	Initiator	Innovation	Award type	Innovation typology
2005	Qingxian County, Cangzhou City, Hebei Province	County government	Village governance model	Finalist	Governance innovation
2005	Maliu Township, Kai County, Chongqing Municipality	Township government	Eight-step work method	Finalist	Management innovation
2005	Daxing District, Beijing Municipality	District government	Women rights defense positions	Finalist	Service innovation
2005	Wuhu City, Anhui Province	City government	Using the Internet to facilitate government–citizen interaction	Finalist	Technological innovation
2005	Yangling District, Xianyang City, Shaanxi Province	District government	Service commitment system	Finalist	Service innovation
2005	Xindu District, Chengdu City, Sichuan Province	District government	The construction of grassroots democratic politics	Finalist	Governance innovation
2005	Jiawang District, Xuzhou City, Jiangsu Province	District government	Public supervision of government affairs	Finalist	Governance innovation
2005	Baixia District, Nanjing City, Jiangsu Province	District government	Huaihai street management system reform	Finalist	Management innovation
2005	Zigui County, Yichang City, Hubei Province	County government	Revoking rural teams and creating communities: new mode of improving villager self-governance	Finalist	Governance innovation
2005	Wuyi County, Jinhua City, Zhejiang Province	County government	Village affairs supervision committee	Finalist	Governance innovation

Appendix 85

2005	Wenzhou City, Zhejiang Province	City government	Efficacy revolution	Finalist	Management innovation
2005	Shaoxing City, Zhejiang Province	City government	The introduction of ISO9000 quality management system in the government office	Finalist	Technological innovation
2005	Changxing County, Huzhou City, Zhejiang Province	County government	Education voucher system	Finalist	Service innovation
2005	Xuhui District, Shanghai City	District government	Reengineering the government process and procedures	Finalist	Management innovation
2007	Xicheng District, Beijing Municipality	District government	Public service hall	Finalist	Service innovation
2007	Huinan Town, Nanhui District, Shanghai City	Township-level people's congress	Public budget system reform	Finalist	Management innovation
2007	Ya'an City, Sichuan Province	City government	Reform of the electoral system of the representatives in township-level people's congress	Finalist	Governance innovation
2007	Sichuan Province	Provincial-level people's congress	Online supervision in budget execution	Finalist	Technological innovation
2007	Jiangxi Province	Provincial-level department of civil affairs	Construction of rural village communities	Finalist	Service innovation
2007	Rui'an City, Wenzhou City, Zhejiang Province	City government	Rural cooperative association	Finalist	Collaborative innovation
2007	Hutubi County, Changji Hui Autonomous Prefecture, Xinjiang Uyghur Autonomous Region	County government	The reform of social pension and insurance system in rural areas	Finalist	Service innovation

(*Continued*)

Table A.1 Continued

Year	Location	Initiator	Innovation	Award type	Innovation typology
2007	Shenzhen City, Guangdong Province	City-level bureau of supervision	The electronic supervision system for administrative licensing	Finalist	Technological innovation
2007	Qingyuan County, Lishui City, Zhejiang Province	County-level organisation department of CCP committee	The construction of skilled township government	Finalist	Service innovation
2007	Yulin City, Guangxi Zhuang Autonomous Region	City government	A "one-stop" new model of e-government	Finalist	Management innovation
2009	Qingyuan Sub-district Office, Daxing District, Beijing Municipality	Township-level office of street affairs	Participatory community governance and community service project management	Finalist	Governance innovation
2009	Xinhe Town, Wenling City, Taizhou City, Zhejiang Province	Township government	Participatory budget reform	Finalist	Governance innovation
2009	Shizhong District, Zaozhuang City, Shandong Province	District-level bureau of finance	Innovation of financial support for agricultural development	Finalist	Service innovation
2009	Meitan County, Zunyi City, Guizhou Province	County-level discipline inspection commission	The concentrated appeal meeting system for villages	Finalist	Governance innovation
2009	Longgang District, Shenzhen City, Guangdong Province	District government	The creation of a comprehensive petition and social stability maintenances system	Finalist	Governance innovation

2009	Huzhou City, Zhejiang Province	City-level organisation department of CCP committee	Innovation of cadre assessment mechanism	Finalist	Management innovation
2009	Jieyang City, Guangdong Province	City-level federation of trade unions	Building trade unions in nonprofit organisations	Finalist	Collaborative innovation
2009	Qingxian County, Cangzhou City, Hebei Province	County government	Construction of rural cooperative pension system	Finalist	Service innovation
2009	Qianjiang District, Chongqing Municipality	District government	Innovation of rural health management system	Finalist	Management innovation
2009	Songyang County, Lishui City, Zhejiang Province	County government	Rural housing sites in exchange for pension	Finalist	Service innovation
2009	Suining City, Sichuan Province	City-level political and legal committee	Social stability risk assessment mechanism	Finalist	Management innovation
2009	Nanjing City, Jiangsu Province	City-level bureau of civil affairs	The registration management system reform for social organisations working for communities	Finalist	Management innovation
2009	Chengdu City, Sichuan Province	City-level federation of trade unions	Inter-provincial trade union contact mechanism for the protection of the rights and interests of migrant workers	Finalist	Service innovation
2009	Nimu County, Lhasa City, Tibet Autonomous Region	County government and CCP committee	Innovation of temple management service system	Finalist	Service innovation
2009	Yanchi County, Wuzhong City, Ningxia Province	County-level external aid project office	Promote public participation in rural communities	Finalist	Governance innovation
2009	Kunming City, Yunnan Province	City-level office of legal affairs	Standardisation of discretionary powers in administrative penalties	Finalist	Service innovation

(*Continued*)

Appendix 87

Table A.1 Continued

Year	Location	Initiator	Innovation	Award type	Innovation typology
2009	Harbin City, Heilongjiang Province	City-level office of legal affairs	Administrative review system reform	Finalist	Management innovation
2009	Huai'an City, Jiangsu Province	City-level bureau of letters and complaints	"Sunshine petition"	Finalist	Service innovation
2009	Pudong New District, Shanghai City	District-level office of the committee of comprehensive management of social security	The mechanism innovation of crime prevention and reduction	Finalist	Service innovation
2009	Liuhe District, Nanjing City, Jiangsu Province	District government	The "Farmer Council" in the natural villages	Finalist	Governance innovation
2011	State Forestry Administration	State Forestry Administration	Reform of collective forest right system	Finalist	Service innovation
2011	China Meteorological Administration, China Meteorological Society	China meteorological administration and China meteorological society	The volunteering program for meteorological disaster prevention and disaster reduction	Finalist	Collaborative innovation
2011	Hainan Province	Provincial-level government service centre	"Three Concentrations" of administrative examination and approval affairs and powers	Special award for media communication	Management innovation
2011	Chengdu City, Sichuan Province	City government	Village-level public service and social management reform	Finalist	Service innovation
2011	Nanchuan District, Chongqing Municipality	District government	The "five non-direct management" delegation reform of party and government leaders' powers	Finalist	Management innovation

Appendix 89

2011	Wenling City, Taizhou City, Zhejiang Province	City government	The collective consultation system of wage	Finalist	Governance innovation
2011	Hangzhou City, Zhejiang Province	City-level office of comprehensive evaluation committee	Citizen-oriented comprehensive evaluation	Finalist	Governance innovation
2011	Baise City, Guangxi Zhuang Autonomous Region	City-level organisation department of CCP committee	Agricultural village office service system	Finalist	Service innovation
2011	Yueqing City, Wenzhou City, Zhejiang	City-level people's congress standing committee	The people's hearing system	Finalist	Governance innovation
2011	Shenzhen City, Guangdong Province	City-level bureau of civil affairs	Privatisation and specialisation of social work	Finalist	Collaborative innovation
2011	Shouguang City, Weifang City, Shandong Province	City government	"Shouguang people's voice"	Finalist	Governance innovation
2011	Anyang City, Henan Province	City-level people's political consultative conference	Deliberation room	Finalist	Governance innovation
2011	Sanxiang Town, Zhongshan City, Guangdong Province	Township-level women's federation	Community integration and development of non-native population	Finalist	Service innovation
2011	Yantian District, Shenzhen City, Guangdong Province	District-level CCP committee	Improve the mechanism of public opinion and promote the construction of grassroots democratic politics	Finalist	Governance innovation
2011	Heping District, Tianjin City	District-level administrative licensing centre	Hiring intermediary organisations in administrative examination and approval services	Finalist	Management innovation

(*Continued*)

Table A.1 Continued

Year	Location	Initiator	Innovation	Award type	Innovation typology
2013	Kailu County, Tongliao City, Inner Mongolia Autonomous Region	County government and CCP committee	Gacha village "532" work approach	Finalist	Service innovation
2013	Ziyang County, Ankang City, Shaanxi Province	County-level CCP committee	A new system for public opinion-oriented cadre selection	Finalist	Governance innovation
2013	Zhaoqing City, Guangdong Province	City-level political and legal committee	"Law of Zhaoqing" Weibo Group	Finalist	Technological innovation
2013	Badahu Subdistrict, South District, Qingdao City, Shandong Province	Township-level office of street affairs	The affiliated model of developing social organisations	Finalist	Collaborative innovation
2013	Chongqing Municipality	Provincial-level financial affairs office	Small loan guarantee and insurance pilot	Finalist	Service innovation
2013	Qingyuan County, Lishui City, Zhejiang Province	County-level discipline inspection commission	The service centre for people in other places	Finalist	Service innovation
2013	Zichuan District, Zibo City, Shandong Province	District-level government audit office	The government directly audits the "village officer" model	Finalist	Management innovation
2013	Taicang City, Suzhou City, Jiangsu Province	City government	Government–society interaction innovation practice	Finalist	Collaborative innovation
2013	Haicang District, Xiamen City, Fujian Province	District government	Social management mechanism innovation of government affairs complex	Finalist	Management innovation

Year	Location	Organization	Project	Status	Type
2013	Beijing municipality	Provincial-level federation of trade unions	The incubation project for employees' services and welfares	Finalist	Service innovation
2015	Guangdong Province	Provincial-level federation of trade unions	"Worker Online" online comprehensive service platform	Finalist	Technological innovation
2015	Yibin City, Sichuan Province	City-level legislative affairs office	Regulations of important political decision-making procedures	Finalist	Governance innovation
2015	Chang Ting County, Longyan City, Fujian Province	County government	"Chang Ting Path" for the comprehensive reform of grassroots healthcare	Finalist	Service innovation
2015	Pengzhou City, Sichuan Province	City-level CCP committee	Grassroots deliberative democracy exploration	Finalist	Governance innovation
2015	Minhang District, Shanghai City	District-level federation of trade unions	Establishment of harmonious capital–labour relations	Finalist	Service innovation
2015	Xiamen City, Fujian Province	City-level health and family planning commission	Reform of the diagnosis and treatment of chronic diseases	Finalist	Service innovation
2015	Xigang District, Dalian City, Liaoning Province	District government and CCP committee	365 Work System	Finalist	Management innovation
2015	Yinchuan City, Ningxia Hui Autonomous Region	City government	The separation of examination and approval powers of the government	Finalist	Management innovation
2015	Wenzhou City, Zhejiang Province	City-level publicity department of the CCP committee	Citizen supervision team	Finalist	Governance innovation

(Continued)

Table A.1 Continued

Year	Location	Initiator	Innovation	Award type	Innovation typology
2015	Gao'an City, Yichun City, Jiangxi Province	City government	Grouped competition, classified assessment, and building an effective government	Finalist	Management innovation
2015	Suzhou City, Jiangsu Province	City-level legislative affairs office	The catalog management of significant administrative decision-making and online operation	Finalist	Management innovation

Note: I classified innovations in the Chinese public sector following Wu, Ma, and Yang (2013), comprising service innovation, technological innovation, management innovation, collaborative innovation, and governance innovation. Specifically, according to Wu, Ma, and Yang (2013), service innovation refers to "the supply of new services to new users, the delivery of existing services to new users or the supply of new services to existing users" (p.350); technological innovation refers to "a change in service delivery technologies or arrangements" (p. 351); management innovation refers to "the restructuring of organisational structures and management processes and practices" (p.351); collaborative innovation refers to "boundary-spanning activities in the process of service delivery and management (for example, alliances, partnerships, collaborations and networking)" (p. 352); governance innovation refers to "new approaches and practices that aim to manage democratic institutions, trigger citizen participation and fight corruption" (p. 352). Notably, many IECLG winners and finalists may involve multiple types of innovation simultaneously. Nevertheless, I classified each IECLG winner or finalist into only one innovation category because the purpose of my classification was only to capture the main characteristic of each IECLG winner or finalist to show the general trend of innovation typology in China.

Index

Page numbers in **bold** denote tables, in *italic* denote figures

administrative: hierarchy 3–4, 17, 22, 24–26, *25–27*, 46–47, 50, 65–69; ladder 22, 24–26, 48, 66; licensing centre (ALC) 9, **12**, **14–15**, 40, **41**, **76**, **86**, **89**

Avellaneda, Claudia N. 1, 7–8, 16, 45, 65, 68

awards programme 3–4, 8, 16, 18, 35–37, **37**, 47–48, 53, 66, 68, **74**

Bai, Guihua 10
Bernier, Luc 16, 46
Berry, Frances S. 1, 7–8, 10, **13**, 16, 24, 31, 45–46, 55, 65, 68

Central Compilation and Translation Bureau (CCTB) 8, 35–36, 53, **54**, **59**, 60
Central Party School of the Chinese Communist Party (CPS) 8, 53, 60
Centre for Chinese Government Innovations at Peking University (CCGI) 35
Chinese Academy of Science (CAS) 22, *28–29*, 38
Chinese Communist Party (CCP) 40, **41–42**, **76–81**, **86–87**, **89–91**
city-level statistical analysis 53, 57, 61
Communist Youth League of China (CYLC) 40, **41**, **80**, **83**
Comparative Politics and Economics Study Centre at the Central Compilation and Translation Bureau (CPESC) 35

Comparative Study Centre of World Parties at the Central Party School of the Chinese Communist Party (CSCWP) 35
Coronavirus Disease 2019 (COVID-19) 38, 69

Deschamps, Carl 16, 46
Duan, Haiyan 3, 17, 25–26

event history analysis (EHA) 10, **11–15**

government: central **13**, **15**, 24, *28–29*, 38, **42**, 50, *51–52*, 53, **54**, 55–57, **56**, **58**; hierarchy 24–25, 51, 68; higher-level 23–26, 28, 48, **59**; innovations 1–4, 7–10, *8–9*, 16, 35, 37, 40, 45–46, 65, 67–68, 70; lower-level 22, 24–26, 28, 30, 48; multi-level 3, 17, 22, 23–24, 27, 67; officials 24, **74**, **85**; provincial 10, **11**, **14**, 16, 25–26, 46, 48, 50, **54**, 55; subordinate 17, 22–23, 30, 50; superior 2, **12**, 17, 22–24, 30, 60–61; system 3, 17, 22–23, *23*, 27, 48, 50, 66
grid governance (GG) 70
Gross Domestic Product (GDP) 30, **54**, 55–57, **56**, **58–59**, 60–62, **61–62**

Hafsi, Taïeb 16, 46
Hasmath, Reza 63, 68

Index

Innovations and Excellence in Chinese Local Governance (IECLG) 3–4, 8, 16, 18, 35–38, **37**, *38*, **39**, 40, **41–42**, 42, **43**, 46–48, **48–49**, 50–53, *51–52*, *54*, **54**, 56–58, **56**, **58–59**, 60–63, **61–62**, 66–68, **74**, **92**
Innovations in American Government Awards (IAGA) 16, 35, 45–46, 68
Innovative Management Award (Canada) (IMA) 16
Institute of Public Administration of Canada (IPAC) 16
intervention policies 9, **14**

Jinping, Xi 37, 63, 68

Keping, Yu 35, 37

land banking system 10, **11**
Landry, Pierre F. 3, 17, 25–26, 30, 66
Li, Wenzhao 10, **12**, 18, 23
Liu, Wei 10, **12**, 18, 23
local government innovativeness 3–4, 7, 17–18, 24, 26–27, 31, 45–47, 50, 52–53, 57, 62, 65–70
Lü, Xiaobo 3, 17, 25–26, 30, 60, 66

Ma, Liang 1, 3–4, 8–10, **11**, **15**, 16, 18, 37, **37**, 40, 45–46, 65, 68, **92**
McGregor, Douglas 27
microblogging 9–10, **11**
Multiple-Plan Integration Reform (MPIR) 4, 68

National Bureau of Statistics (NBS) **54**, 55, **59**, 60
National Forestry Administration (NFA) **42**, **48**
National Meteorological Administration (NMA) **42**, **48**
New Public Management (NPM) 1, **13**

Peking University (PKU) 8, 35–36, 53, **54**, **59**, 60
pension policies 10, **15**
performance management 1, 10, **12**, 65, **78**
Poisson Autoregression Regression (PAR) **14**
policy: innovations 7, **11**, **14**, 50; instruments **12**, 23, 25; -maker 2, 9, *9*, 35, 50, 65, 68

population **54**, **56**, **58–59**, **61–62**, 63, **89**; size **54**, 55, 57, **59**, 60–61, 63
Province-Managing-County (PMC) 4, 50–51, 68
provincial-level statistical analysis 53
Public Administration (PA) 2–3, 7, 10, **11–15**, 36
public policy (PP) 2–3, 7, 10, **11**, 36
public–private partnership 1, 10, **12**, 65

Research Centre for Chinese Politics (RCCP) 36
Resource and Environment Data Cloud Platform (REDCP) 22, *28–29*, *38*

science: management 3, 7, 17; political 3, 7, 36
socioeconomic development 1–2, 8, 67, 69
span of control 3–4, 17, 22, 27, 30–31, 50–53, 55, 57, 60, 62, 65–69
statistical analysis 3, 53, 56–57, 61, 66–67
Super-Department Reform (SDR) 4, 68

Teets, Jessica C. 63, 68
Turning the Prefectures into Cities Reform (TPCR) 4, 68

United States 2, 7, 31, 46, 68, 70
Urban Minimum Living Standard Assistance System (UMLSAS) 10, **13**

Walker, Richard M. 1, 7–8, 16, 45, 65, 68
Wu, Jiannan 1, 3, 8–10, **13–14**, 18, 37, **37**, 40, 46, **92**

Yang, Yuqian 3, 8, 18, 37, **37**, 40, 46, **92**

Zhang, Pan 1, 9–10, **13–14**, 23
Zhang, Yanlong 1, 8, 10, **11–12**, **54**, 55, **59**, 60
Zhang, Youlang 1, 3, 8–10, **12**, **14–15**, 22–25, 27, 30–31, 55–56, 61
Zhao, Hui 10, **13**, **15**, 18
Zhao, Qiang 45
Zhu, Xufeng 1, 3, 8–10, **12–15**, 18, 22–27, 30–31, 40, 53, 55–56, 61, 66
Zhu, Yapeng 9–10, **11**, 31